This book is dedicated to those who appreciate a mathematical challenge, or those who delight in a puzzle or riddle, or those who may enjoy learning about something unfamiliar, and especially to those who may be indifferent to all of the above.

I0478322

PREFACE

This book is comprised of three sections.
Section 1 introduces Mathematical challenges that include the concepts of algebra, geometry, trigonometry, probability, and logic.
Section 2 offers a mixture of Puzzles and Riddles to stimulate the imagination.
Section 3 is a collection of Expositions that convey information or explanation about various topics.

Appendix 1 provides the answers to the Section 1 Mathematics questions.
Appendix 2 provides the answers to the Section 2 Puzzles and Riddles.

There is <u>no</u> **Appendix 3**, because the Expositions of Section 3 present explanatory details and information with no questions for the reader to answer.

Clearwater, Florida
March, 2015

Merle

QUESTIONS
???????

P
R
O
B
L
E
M
S

Mathematics

Puzzles

Riddles

Expositions

S
O
L
U
T
I
O
N
S

!!!!!!!
ANSWERS

Merle A. Barlow

WestBow Press books may be ordered through booksellers or by contacting:

WestBow Press
A Division of Thomas Nelson & Zondervan
1663 Liberty Drive
Bloomington, IN 47403
www.westbowpress.com
1 (866) 928-1240

ISBN: 978-1-4908-6566-9 (sc)
ISBN: 978-1-4908-6568-3 (hc)
ISBN: 978-1-4908-6567-6 (e)

Library of Congress Control Number: 2015900220

Printed in the United States of America.

WestBow Press rev. date: 01/26/15

WESTBOW·
P R E S S
A DIVISION OF THOMAS NELSON
& ZONDERVAN

Prologue

Mathematics, rightly viewed, possesses not only truth, but supreme beauty – a beauty cold and austere, like that of sculpture, without appeal to any part of our weaker nature, without the gorgeous trappings of painting and music, yet sublimely pure, and capable of a stern perfection such as only the greatest art can show.

Bertrand Russell

Philosophy is written in this grand book – I mean the universe - which stands continually open to our gaze, but it cannot be understood unless one first learns to comprehend the language and interpret the characters in which it is written. It is written in the language of mathematics, and its characters are triangles, circles, and other geometrical figures, without which it is humanly impossible to understand a single word of it; without these, one is wandering about in a dark labyrinth.

Galileo

God in the beginning formed matter in solid, massy, hard, impenetrable, moveable particles, of such sizes and figures, and with such other properties, and in such proportion to space, so as to appropriately contribute to the end for which He formed them.

Sir Isaac Newton

...(In Christ)...are hidden all the treasures of wisdom and knowledge.

Colossians 2:3

CONTENTS

Section 1
Mathematics

CONTENTS

Section 2
Puzzles & Riddles

CONTENTS

CONTENTS

CONTENTS

**Section 3
Expositions**

CONTENTS

SECTION 1
MATHEMATICS

1
The Bookworm
An Arithmetic & Logic Problem

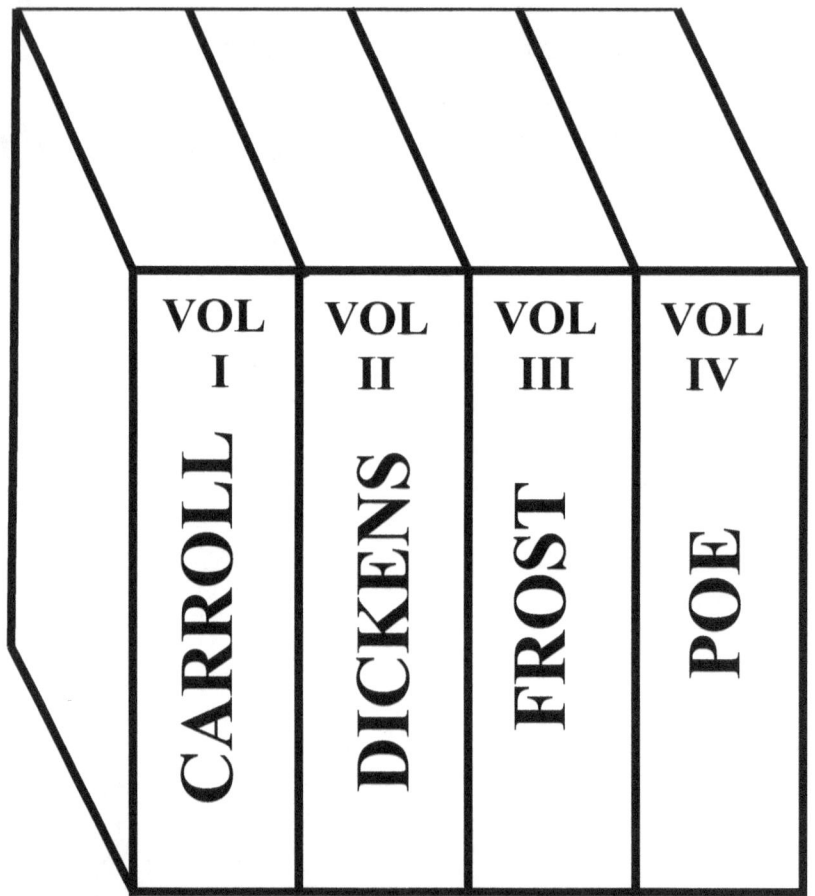

[In a library, four volumes of literature are shelved as illustrated in the diagram above.]

An erudite worm starts at the beginning of Volume I and eats his way to the end of Volume IV. If each of the covers is 1/8 of an inch thick, and the pages of each book have a total thickness of 2 inches, how far has this intelligent worm eaten?

Solution Option:

Instead of arithmetic and logic, you can solve the calculus integral: The numerical value of this integral is equal to the

number of inches eaten by the worm. $\int_{0}^{\sqrt[4]{19}} x^3 \, dx$

2
A Logic Problem For The Ages

Question:
There is only one time in your life when you are twice as old as your child. When is that?

3
Right Triangle Relationships

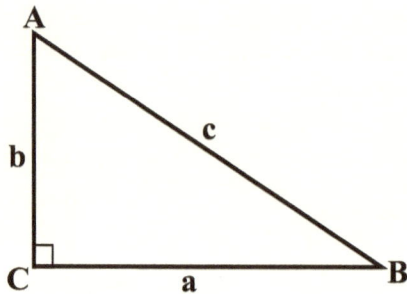

Consider the right triangle ABC with right angle C.
By the Pythagorean Theorem,

$c^2 = a^2 + b^2$, and
$c = \sqrt{a^2 + b^2}$

Question:
What is the relationship between (a + b) and c?
In other words, define this relationship mathematically.

Hint: There are exactly three possibilities, and only one is true.

4
Thinking Outside the Box

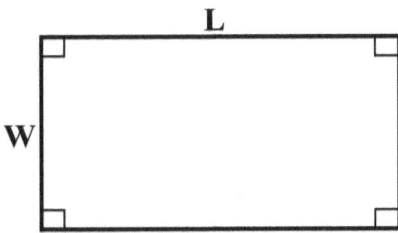

The area of a rectangle is usually expressed in terms of its length (L) and width (W). My challenge to you is to consider the unusual:

(I) Develop an equation to express the area of a rectangle, but do not use the width variable (W).

$$A = ?$$

(II) Develop an equation to express the area of a rectangle, but do not use the length variable (L).

$$A = ?$$

5

Thinking Outside the Circle

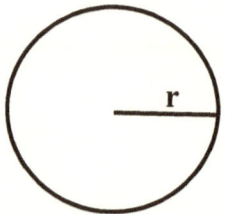

The circumference and area of a circle are usually expressed in terms of its radius or diameter. My challenge to you is to confront the extraordinary:

(I) Develop an equation that expresses the circumference of a circle in terms of the circle's <u>area</u>. In other words, the equation will <u>not</u> contain a variable for the radius or diameter.

$$C = ?$$

(II) Develop an equation that expresses the area of a circle in terms of the circle's <u>circumference</u>. In other words, the equation will <u>not</u> contain a variable for the radius or diameter.

$$A = ?$$

6
Head Travel

Suppose you went on a long walking tour around the earth's equator, and suppose that you are six feet tall. (For the purpose of friendlier numbers for calculations, assume that the earth is a perfect sphere and the distance around the equator is 25,000 miles.)

Question:
How much farther would your head travel than your feet?

7
Traveling in Circles

Assume that two cars are both traveling with speeds of 70 mph on roads around the equator of a perfectly spherical earth whose diameter is 8000 miles. The red car is traveling on a road directly on the equator; the blue car is traveling on a road elevated 15 feet above the equator for the entire distance around the earth.

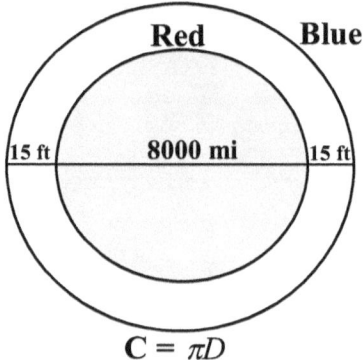

Question:
How much longer will be required for the blue car to finish one revolution?

8
Cubical Water Tank

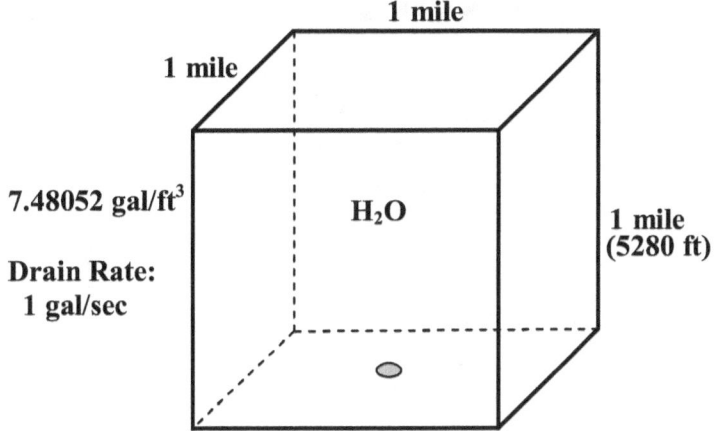

A cubical water tank has an edge equal to 1 mile (5280 ft.), and is completely filled with water. In the middle of the bottom side of the tank is a drain plug. If opened, the water will drain at the rate of 1 gallon per second.

Question:

What time will be required to completely empty the tank? Express your answer to the nearest whole year.

9
A COIN PROBABILITY PROBLEM

A box contains two coins. One coin has "heads" on both sides. The other coin has "heads" on one side and "tails" on the other side. All three "heads" have <u>identical</u> configurations. A coin is selected randomly. The observed side of the selected coin is "heads".

Question:
What is the probability that the other side of the selected coin is <u>also</u> "heads"?

10
A LOGIC/PROBABILITY PROBLEM

Four identical sealed envelopes are on a table. One of them contains a $100 bill. You randomly select an envelope and hold it in your hand without opening it.

Two of the three remaining envelopes are then removed and set aside, unopened. You are told that these two envelopes are empty (they truthfully are empty).

You are given the choice of keeping the envelope you chose or exchanging it for the one on the table.

What should you do?
      (A) Keep your envelope
      (B) Switch it
      (C) It does not matter

11
PROOF THAT 1 = 2

1. *Let A = B*

2. Square both sides of the equation (Multiply both sides by *B*):
    $$(A)(B) = B^2$$

3. Subtract A^2 from both sides of the equation:
    $$(A)(B) - A^2 = B^2 - A^2$$

4. Factor both sides of the equation:
    $$A(B - A) = (B + A)(B - A)$$

5. Divide both sides of the equation by $(B - A)$:
    $$A = (B + A)$$

6. By the authority of step #1, if $A = 1$, *then B = 1*:
    $$1 = 1 + 1$$

7. Therefore, $1 = 2$

Q.E.D.

Postscript:
Can you determine the error in this proof, or does your indifferent sensibility result in your contentment to accept its invalidity?

12
Inscribed & Circumscribed

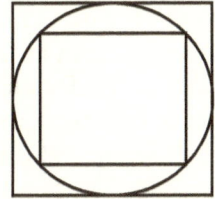

In the illustration above:

From an inside to outside perspective:
The small square is inscribed in the circle.
The circle is inscribed in the large square.
The larger square has an area of 10 square units.

From an outside to inside perspective:
The larger square is circumscribed about the circle.
The circle is circumscribed about the smaller square.
The larger square has an area of 10 square units.

Question:
What is the area of the smaller square?

13
Tissue Paper Thinking

Imagine a steel band fitting tightly around the equator of the earth (assume that the earth is a perfect sphere). Now, suppose that you remove it and then splice in an additional piece 10 feet long, so that the new band is 10 feet longer than the original band. If you now replace it on the equator, it would fit more loosely all around the circumference of the earth.

Question:

How great a distance would there now be between the band and the earth?

Would the distance be sufficient for:

(a) A person, 6 feet tall, to walk through?

(b) A person to crawl through on hands and knees?

(c) A piece of tissue paper to just slip through?

(d) None of the above?

14
An Algebra Problem With A Temperature

At high noon on October 16, 2012 in Clearwater, Florida and in Lafayette, Indiana, the sum of their 2-digit Fahrenheit temperatures was 121 degrees.

The tens' digit of the Clearwater temperature was three greater than its units' digit. If you reversed the two digits of the Clearwater temperature, the result was the Lafayette temperature.

Question:
What were the temperatures of these two cities?

Anyone can guess at anything; however, for this problem:
You do the math!
        and
You show the math!

15
A CARD PROBABILITY PROBLEM

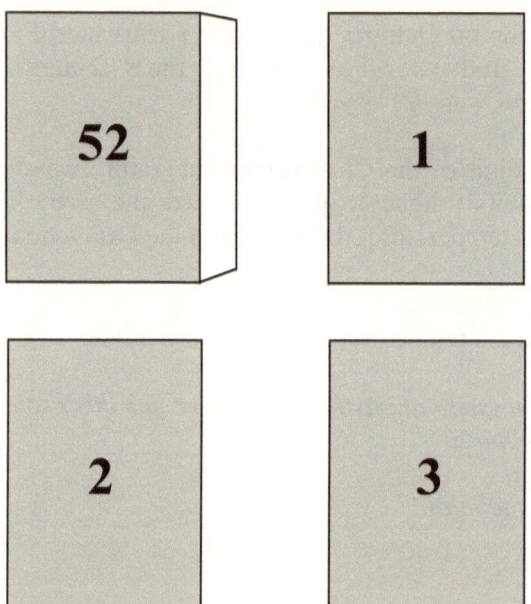

An ordinary deck of 52 playing cards (already thoroughly shuffled) is shuffled completely (this means the deck is shuffled and repeatedly reshuffled, cut, re-cut, reshuffled, cut, re-cut, etc.). You may assume that the card deck is in a "theoretical random" sequence. That is, the sequence is not biased in any manner. Three cards are sequentially removed from the top of the deck and placed face down. Card 1 is removed first; Card 2 is the second card removed; Card 3 is the third card removed.

There are four probability questions. Express each of the four probabilities as a Decimal Fraction <u>rounded</u> to four decimal places (e.g., .XXXX), and as a Per Cent <u>rounded</u> to two decimal places (e.g., XX.XX%).

(1) What is the probability that NONE of the three cards is a Red Queen?

Suppose that <u>only</u> <u>one</u> of these three cards IS a Red Queen. <u>Which</u> <u>card</u> is <u>most</u> <u>probably</u> the Red Queen? The answer can be determined after answering the next three questions.

(2) What is the probability that the FIRST card is a Red Queen?

(3) What is the probability that the FIRST card is NOT a Red Queen, AND the SECOND card IS a Red Queen?

(4) What is the probability that the FIRST card is NOT a Red Queen, AND the SECOND card is NOT a Red Queen, AND the third card IS a Red Queen?

16
Poetic Equation

Question:
Can you read the following equation as a poem?

$$\frac{12 + 144 + 20 + 3\sqrt{4}}{7} + (5 \times 11) = 9^2 + 0$$

17
A MARBLES LOGIC PROBLEM

An urn contains exactly 100 spherical glass marbles having the same smooth texture. There are five different colors. The color distribution is as follows:

    30 Blue marbles

    30 Red marbles

    30 Green marbles

    The 10 remaining marbles are an unknown mixture of White and Yellow.

The urn is shaken thoroughly and a marble is selected unseen. This procedure is repeated for each marble selected.

Question:

What is the <u>Absolute</u> <u>Minimum</u> number of marbles that must be selected in order to <u>Assure</u> with <u>Absolute</u> <u>Certainty</u> that 10 marbles of the <u>same</u> color have been selected?

18
DIE PROBABILITY PROBLEM

Question:

Suppose you plan an experiment in which you will roll a die 20 times. Which of the following results is more likely?

    (A) 11111111111111111111

    OR

    (B) 66234441536125563152

19
Basket of Eggs

A country grocery store clerk prepares a basket of fresh eggs ordered and purchased by a customer. The customer comes to the store and tells the clerk:

"I'd like half the eggs in the basket plus half an egg, please."

The clerk gives the requested number of eggs to the customer. The following day, the customer returns and instructs the clerk:

"I'd like half the eggs in the basket plus half an egg, please."

Again the clerk provides the requested number of eggs for the customer. The next day. the customer returns again and tells the clerk:

"I'd like half the eggs in the basket plus half an egg, please."

The clerk gives the requested number of eggs to the customer, and the basket is now <u>empty</u>.

Question:
How many eggs were originally in the basket?

20
Sphere in a Cube

Given:

A sphere is inscribed in a cube (which also means that the cube is circumscribed about the sphere).

The diagonal (AB) of the cube face = $3\sqrt{6}$ *meters*.

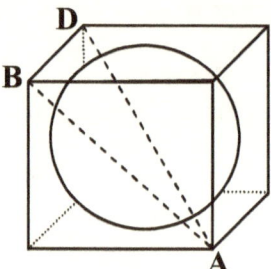

Use a calculator for the calculations involved in the following questions. There are ten questions. Each question is worth ten points. The total number of points is 100. 100% represents Success. ☺; any other percentage is Failure. ☹

What is the volume of the sphere?

> ➢ Use the symbol "π" as π in the answer. Do not convert π to any numerical value. Simplify the expression of the answer as much as possible. If any part of the answer cannot be converted to a rational number, leave it as a radical (do <u>not</u> convert the radical to a numerical value).

(1) _____   (2) _____
    (calculated value)   (unit of measurement)
                            (same for (1) and (3))

> ➢ Use the calculator value of π, completely calculate the expression of (1) including any radical, and round the completely numerical answer to <u>two</u> decimal places.

(3) _____
    (calculated value)

22

What is the ratio of the spherical volume to the cubical volume?

> ➤ Use the calculator value of π, round the answer to <u>two</u> decimal places, and multiply by 100 to express the answer as a percent.

(4) _____%
(calculated percent)

What is the ratio of the surface area of the sphere to the surface area of the cube?

> ➤ Use the calculator value of π, round the answer to two decimal places, and multiply by 100 to express the answer as a percent.

(5) _____%
(calculated percent)

What is the circumference of the sphere?

> ➤ A plane passed thru the center of the sphere describes a great circle on the sphere's surface. This circle represents the perimeter or circumference of the sphere.

> ➤ Use the calculator value of π, and round the answer to <u>two</u> decimal places.

(6) _____ *meters*
(calculated value)

What is the area of the planar area ABD?

> ➤ This triangular area is formed by lines AB, BD, and AD.

> ➤ Round the answer to <u>two</u> decimal places.

(7) _____ (8) _____
(calculated value)   (unit of measurement)

What is the length of the cube diagonal AD?

➤ Round the answer to <u>two</u> decimal places.

(9) _____ *meters*
      (calculated value)

How many points of tangency (contact points) are there between the sphere and the cube?

➤ Express the answer as an integer.

(10) _____

21
Precision Design

As illustrated in Diagram 1, a perfect sphere is inscribed in a perfect cube. The dimensions of both are precisely designed so that the sphere and cube are tangent to each other. This indicates that the sphere and cube contact each other at only six points – the exact centers of each cube face. An edge of the cube has a length of "D". A second, smaller sphere, is positioned in the front right corner of the cube. This sphere is tangent to the large sphere, and tangent to the cube on the right, front, and bottom faces. This indicates that the small sphere contacts the large sphere at only one point, and the small sphere contacts three cube faces – right, front, and bottom, for a total of four contact points. The cube's diagonal from the top left rear vertex to the bottom right front vertex, intersects the centers of each sphere, and passes through the tangential contact of the two spheres.

As illustrated by Diagram 2, the small sphere is inscribed in a small cube such that both are tangent to each other. An edge of the small cube has a length of "d". The small sphere is the same sphere that is illustrated in Diagram 1. Obviously, the small cube of Diagram 2 cannot possibly reside in the large cube. It is illustrated here to define the dimension of the small sphere, and to illustrate the fact that the small cube diagonal is indeed a segment of the large cube diagonal.

The Challenge

(1) Determine the equation that expresses the small sphere diameter in terms of the large sphere diameter.

(2) After determining the appropriate equation, find the value of the small sphere diameter to the nearest millionth of a unit (six decimal places) when the large sphere diameter is equal to 1 unit.

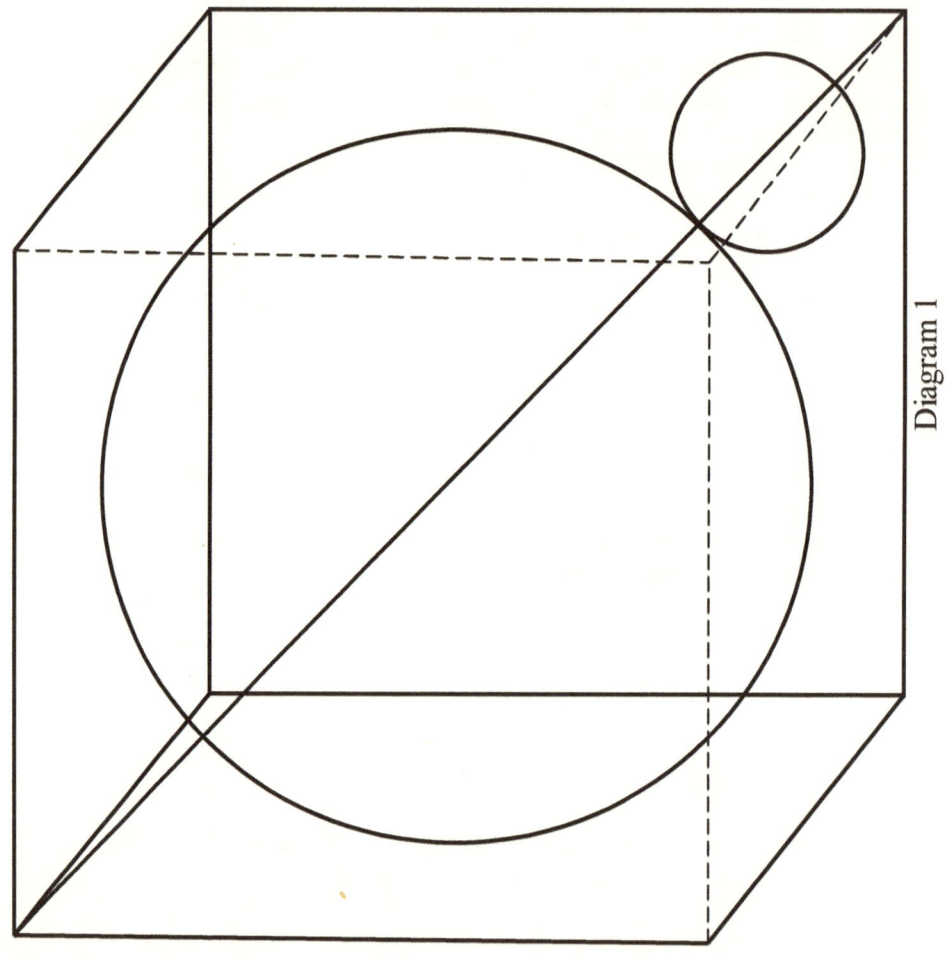

Diagram 1

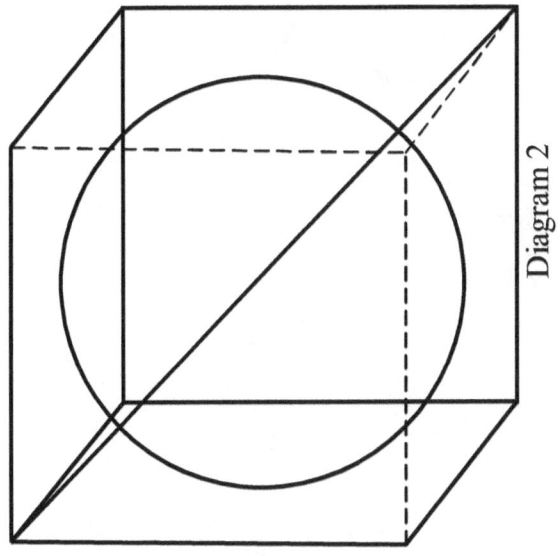

Diagram 2

Algebra – Fundamental Operations
EXAM

Simplify (a), (b), (c), and (d).
Then perform the operations specified by (e), (f), and (g).
All variables are alphabetic characters.
After operation (g), <u>rearrange the eleven alphabetic characters</u> to form a three-word phrase. ☺

(a)
$$\frac{J\left[\left(E^{\frac{1}{2}}R^{\frac{1}{2}}Z - A\right)\left(E^{\frac{1}{2}}R^{\frac{1}{2}}Z + A\right)(S)^2 + S^2 A^2\right]}{RSZ^2}$$

(b)
$$\frac{\dfrac{U+S}{US} + \dfrac{S}{\dfrac{U^2 - US}{U}}}{S^2\left(U^2 - US\right)}$$

(c)
$$EY\left(\frac{4IS}{3\left((E+Y)^2 - (E-Y)^2\right)}\right) + ZY\left(\frac{4IS}{3\left((Y+Z)^2 - (Y-Z)^2\right)}\right)$$
$$+ EZ\left(\frac{4IS}{3\left((Z+E)^2 - (Z-E)^2\right)}\right)$$

(d)
$$\frac{2L\sqrt{C^2 Z^2 \sqrt[3]{O^6}}}{3BZ} \times \frac{3\sqrt[3]{B^3 Z^3 \sqrt{(RD)^6}}}{2CZ}$$

(e) Multiply (a) and (b).

(f) Add (c) to (e).

(g) Add (d) to (f).

23
Mathematics – Fundamental Algebra
EXAM

(1)
Perform the indicated operations and simplify:

$$\frac{\dfrac{a-b}{a}-\dfrac{b}{a+b}}{\dfrac{a+b}{b}-\dfrac{a}{a-b}}+\frac{1-\dfrac{b}{a}}{\dfrac{a}{b}+1}$$

(2)
Solve the system of simultaneous linear equations for x, y, and z:

$$\frac{3x}{2}-y+\frac{2z}{3}=4$$

$$\frac{x}{3}+\frac{5y}{2}-\frac{z}{6}=\frac{8}{3}$$

$$\frac{x}{6}-\frac{2y}{3}+\frac{z}{9}=0$$

(3)
Perform the indicated operations and simplify:

$$\frac{y^{-2}\left(x^{3}y^{-1}+x^{2}\right)}{x^{2}\left(1+x^{-1}y\right)}$$

(4)
Solve the inequality for x, then graph the solution set:

$$\left(\frac{x+8}{4}-1\right)>\frac{x}{3}$$

(5)

In aerospace engineering, the radius (r) of a satellite orbit is determined by the formula below, where t represents the time required for the satellite to complete one orbit, G represents the universal gravitational constant, and M represents the mass of the central body (earth). Solve the formula for t:

$$r = \sqrt[3]{\frac{GMt^2}{4\pi^2}}$$

(6)

The following proof is presented for your analysis. The conclusion, obviously wrong, is logically consistent from the sequence of mathematical operations; however, one of the steps is <u>invalid</u>. Your objective is to determine which step (other than step 7) is invalid, and why.

PROOF THAT 1 = 2

1. Let $A = B$
2. Square both sides of the equation:
   $$AB = B^2$$
3. Subtract A^2 from both sides of the equation:
   $$AB - A^2 = B^2 - A^2$$
4. Factor both sides of the equation:
   $$A(B - A) = (B + A)(B - A)$$
5. Divide both sides of the equation by $(B - A)$:
   $$A = B + A$$
6. From Step 1, if $A = 1$, then $B = 1$:
   $$1 = 1 + 1$$
7. Therefore, $1 = 2$

Question 1: Which step is invalid? _____
Question 2: Why is the step invalid?

_____

(7)

According to current Florida State Law, the sales tax on a new automobile is calculated as follows:

The first $5000 of the sale price is taxed at the rate of 7%.

The balance of the sale price is taxed at the rate of 6%.

Let S represent the Sale price of an automobile. Write the equation, in terms of S, to calculate the Total Sale price (including the tax). Express the equation in simplest form. Do not substitute some specific numerical value for S. The objective is to determine an algebraic equation that is a general solution.

(8)

Assuming that fractional hens are as proportionally functional as whole hens, consider the following question:

If a hen-and-a-half lay an egg-and-a-half in a day-and-a-half, how many eggs will three hens lay in three-and-a-half days?

To avoid possible confusion, the same question will be expressed differently:

If 1.5 hens lay 1.5 eggs in 1.5 days, how many eggs will 3 hens lay in 3.5 days?

Hens	Eggs	Days
3/2	3/2	3/2
3	?	7/2

(9)

Suppose the earth is a perfect sphere, and there is a steel band fitting tightly around the equator. Now suppose that you remove the band and cut it at one place, then add an additional piece that is 10 feet long. The new band is now 10 feet longer than the original band. If you replace the band on the equator, it will fit more loosely. It will fit more loosely because an increased circumference also increases the radius. Previously, there was no space between the band and the earth, but now the increased radius causes a space between the band and the earth all around the equator. (Reference the diagram).

Let C = circumference of the original band in feet.

Let R = radius of the original band in feet.

Let r = the increase of the radius in feet (only the increase, not the entire radius of the new band).

Let π = 3.14

There are two questions:

(1) What is r, to the nearest <u>tenth</u> of a foot?

(2) Select the answer below that is most logically consistent with the value of r.

> The space (r) between the new band and the earth will be great enough for:
>
> ____(a) a person, 6 feet tall, to walk through, or
> ____(b) a person to crawl through, or
> ____(c) a piece of tissue paper to tightly slip through, or
> ____(d) none of the above?

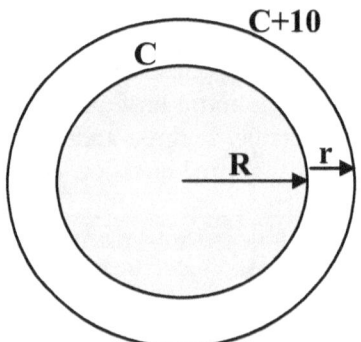

(10)

Suppose that a singles tennis tournament has been planned. It will be a double-elimination tournament (a participant is eliminated after losing twice). The number of participants will be N. One can of tennis balls will be provided for each and every tennis match required. Expressed in terms of N, how many cans of tennis balls are necessary for this tournament? (Because the number of participants is N, do not express the answer with a specific numerical value for the number of cans. The answer is an algebraic expression for a general solution).

24
Up Against A Brick Wall 1

A brick wall (wall #1) is one brick higher and one brick narrower than another wall (wall #2). **How many <u>more</u> bricks are in wall #1?** [Assume that all the bricks are square].
You can obviously count the bricks in the illustration.
Develop an algebraic formula to prove the general solution.

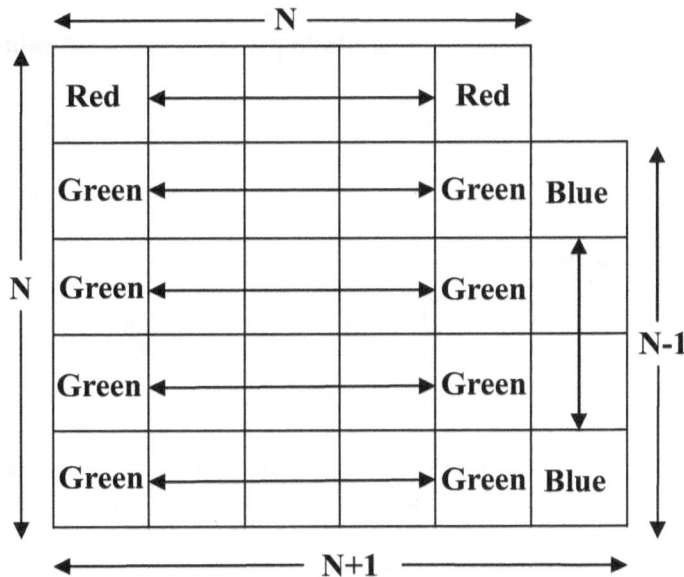

The square bricks that are Red and Green represent the square Wall #1.
The square bricks that are Green and Blue represent Wall #2.
The Green bricks are common to both walls.

25
Up Against A Brick Wall 2

Consider an alternative problem regarding the two brick walls having the condition that all bricks are rectangular (not square). This is actually the normal situation. The dimensions of height (h) and width (w) of each brick are unequal, but all the bricks are exactly the same size.

Now, how many bricks does one wall have relative to the other wall?
Develop an algebraic formula to prove the general solution.

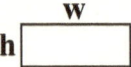

w > h for each brick; The front surface area of each brick = hw

Bricks in Wall #1 are Red and Green (the same color pattern as "Up Against A Brick Wall 1").
Bricks in Wall #2 are Blue and Green (the same color pattern as "Up Against A Brick Wall 1").
The Green bricks are common to both walls.

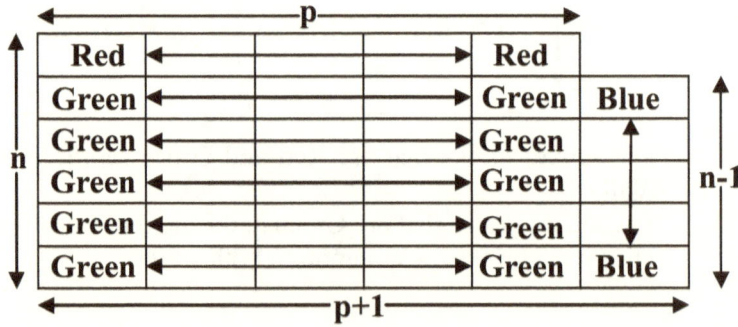

26
Box of Balls

Question:
What is the maximum number of one-inch diameter, non-compressible stainless steel spherical balls that can be packed into a cubical stainless steel box that is 100 inches on the inside edge?
(The box top must close completely).

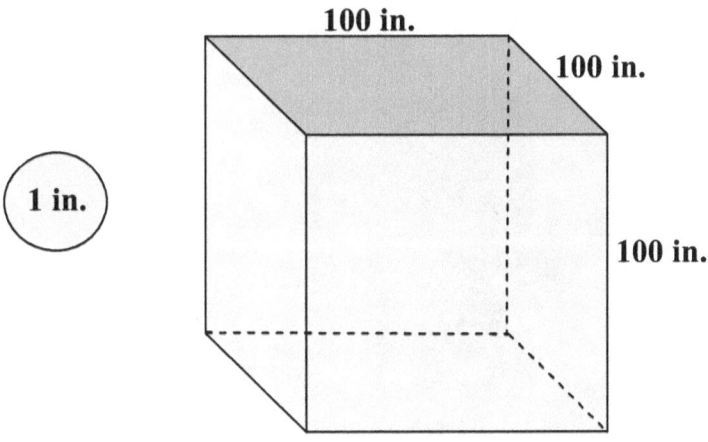

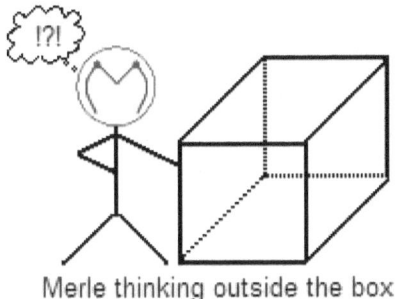

Merle thinking outside the box

27
A Number Base Problem

Solve for x:

$324_x = 3064_9$ [324 (base x) = 3064 (base 9)]

(Hint: A quadratic equation is an efficient method of solution.)

SECTION 2
PUZZLES & RIDDLES

Puzzles & Riddles

1 Fragility
What is so fragile that when you say its name, you break it?
? ? ? ? ? ? ? ? ? ? ? ? N ?

2 Forward & Backward
Forwards, I am heavy; backwards, I am not. What am I?
? ?

3 Emotion Malady
Question:
Is it any wonder why so many people in this world experience the alternation of manic and depressive states of emotion?
Do you know why this is true?
? ?

4 A Fair-weather Friend
I come in only one color,
But numerous shapes you will see.
I stay in the sun, but not the rain.
No one can ever harm me,
And I will never feel pain.
What am I?
? ?

5 A Convenient Accessory
Solve the following riddle for something that everyone has seen:

It begins with a letter found only once in this sentence.
You see it infrequently in the sun.
It comes in many different colors, but only one general shape.
A food that is eaten resembles its shape.

What is this something?
? ?

6 A Delicious Word
In reference to the sequence of letters below:
By crossing out six letters, the remaining letters, in sequence, spell a common word.
What is the word?

BSAINXLEATNTEARS
? ?

7 A Special Number
The following number is extraordinarily special:
8,549,176,320
Do you know why?

8 Western State Capitals
How many U.S. state capitals are <u>west</u> of Los Angeles?
? ?

9 Logically a Poor Fit
Which of the following items does not belong with the others:
Binoculars; Eyeglasses; Goggles; Handlebars; Jeans; Pliers;
Scissors; Shoes; Tweezers?
? ?

10 Animal Arithmetic
What do you get when you add the following?:
3/7 chicken plus 2/3 cat plus 1/2 goat
? ?

11 Month After Month
This is a challenge:
What are the missing numbers in this series?:
4   4   1   2   1   1   2   2   3   (?)   (?)   (?)
(Hint: The quantity of numbers in this series is a significant
hint.)
? ?

12 You Are Looking at the Answer
Which letter comes next in this series?:
W L C N I T (?)
? ?

13 This is Difficult and Puzzling
What is the missing number in the following sequence?:
20, 22, 24, 26, 30, 33, 40, 44, 120, ? , 11000
? ?

14 A Chemical Compound
What common substance is represented by the following letters?:
HIJKLMNO
? ?

15 Irish Eyes are Smiling
In regard to St. Patrick's Day, here is an interesting riddle:
Who Is Irish And Stays Out In The Sun All Day?
? ?

16 An Alphabetic State of Reference
This is a difficult sequence puzzle:
What three numbers complete this sequence?:
4031011140218831012212 4???
? ?

17 A Musical Mixture
How are the following words logically related?:
Dome, Reign, Mink, Father, Sodium, Lava, Tingle, Dough
? ?

18 A Sneaky Riddle
How do you catch a **rare** bird?

? ?

19 Number Fashions
What did the zero say to the eight?

? ?

20 Number Fears
Question:
Why was 6 afraid of 7?

? ?

21 An Order Perspective
Can you decipher this sentence?:

Eht eelrstt fo aceh dorw in hist ceeennst aer in aaabcehillpt deorr.

? ?

22 The Great Sahara
The Sahara Desert is the world's largest desert (3,500,000 square miles).

Question:
Why can you <u>not</u> starve in the Sahara Desert?

? ?

23 Homophones
(The following Riddle must be presented orally, not written, as will be evident).
Question:
One Knight, a Biologist, a Doctor, and Big Foot got in a boat and went out to the middle of the lake where the Loch Ness Monster lived. The Monster first ate the Doctor, then he ate the Biologist, and then Big Foot. Who was left in the boat?

? ?

24 Be Enlightened
I am tall when I am young.
I am short when I am old.
When I live, I glow.
From your breath, I die.

What am I?

? ?

25 Logical Progression
Question:
Following are the first letters of six words:
L, W, S, P, C, B
If the sixth word is "Book", what are the first five words?

? ?

26 A Clever, but Difficult Series
Can you write the next line in the following infinite series of lines?

1
11
21
1211
111221
    ?

? ?

27 Falling and Breaking
Question:
What falls, but never breaks, and what breaks, but never falls? (two separate things)

? ?

28 Fore!
Question:
Suppose that four golfers drive <u>identical</u> balls onto a green, where they land close to one another. No one knows which ball is whose, so each player randomly chooses a ball. What are the chances of <u>exactly</u> three of the four golfers selecting their own balls?

? ?

29 A Courtly Group
Question:
What category of people live in a world where receiving fast service is <u>not</u> appreciated, and someone's faults are someone else's successes, and love means nothing at all?

? ?

30 An Illustration of Relativity
What did the Snail exclaim while riding on the back of the Turtle?

? ?

31 A Bucket of Mystery
There is something that you can see, and if you put it in a bucket, the bucket will be lighter.
What is it?

? ?

32 History/Monroe Doctrine
There is a three-word phrase that correctly and explicitly defines the Monroe Doctrine.
What are the three words?

? ?

33 A Corny NFL Riddle

The Tampa Bay Buccaneers defeated the Oakland Raiders to become Super Bowl XXXVII Champions. Because of their victory, the market price ot corn in the Tampa Bay area was elevated to a record high.

What was this new record price for corn?

? ?

34 Sesquipedalianism

There is a respiratory ailment that is named "Black Lung" disease. This name is a shortened form of what is recognized as the longest word in the English Language.

Are you familiar with this word? (It has only 45 letters!).

? ?

35 Commonality

Kermit and John the Baptist have something in common.
What is it?

? ?

36 Colors

What is Blue, but smells like Red Paint?

? ?

37 Buses and Busses

How are Buses (transportation) and Busses (kisses) logically related? [other than spelling]

? ?

38 Old Testament Facts

What was the name of the Prophet Jeremiah's Horse?

(Jeremiah referred to "horsemen" in the Fourth Chapter. Following this reference, he mentioned his horse's name three times in the Book of Jeremiah. Refer to a KJV or NKJV Bible).

? ?

39 A Toothpick Cat

Puzzle:

On a piece of paper, nine toothpicks are arranged as follows to form 100:

The challenge is to alter the position of only <u>two</u> toothpicks to spell the word CAT.

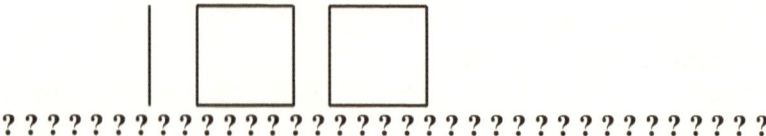

? ?

BLACK

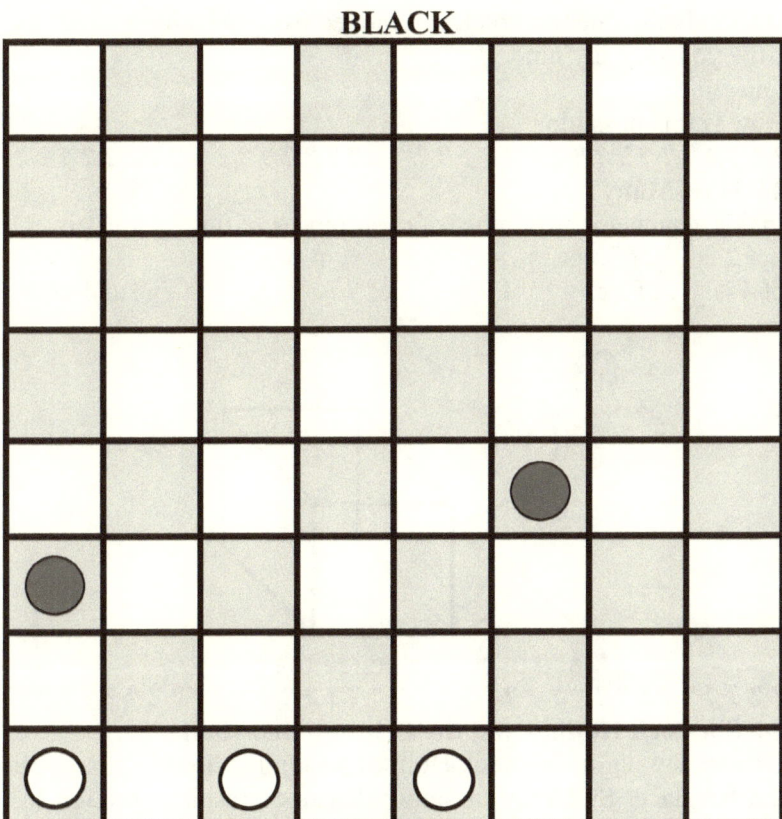

WHITE

(Black's Move)
White declares that Black will not be able to crown the piece he moves first.
Is White correct?

? ?

41 A Puzzling Picture
What is it?

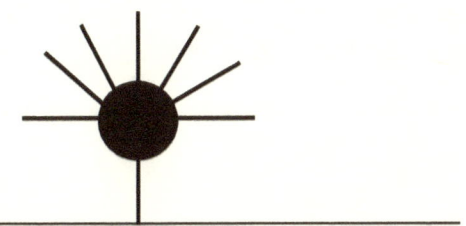

? ?

42 Twelve to Nine
There are twelve toothpicks on a table.
One and only one toothpick is removed from the table.
Now you see only nine.
Question:
How is this possible?
? ?

43 How Many?
In this sentence, the number of occurrences of 0 is __ , of 1 is __ , of 2 is __ , of 3 is __ ,
of 4 is__ , of 5 is __ , of 6 is __ , of 7 is __ , of 8 is __ , and of 9 is __ .

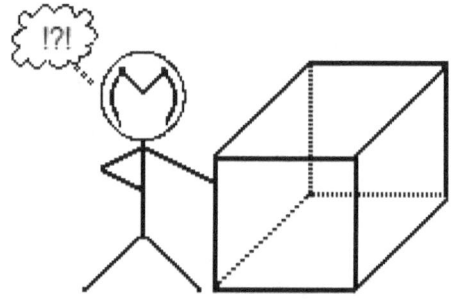

Merle thinking outside the box
? ?

44 Sherlock Holmes and the Crimson Snowball
One snowy night, Sherlock Holmes was in his house sitting near the fireplace. Suddenly, a snowball came crashing through his window, breaking it. Holmes got up and looked out the window just in time to see three neighborhood kids who were brothers run around a corner. Their names were John Crimson, Mark Crimson, and Paul Crimson. The next day, Holmes got a note on his door that read, "? Crimson. He broke your window."
Which of the three Crimson brothers should Holmes question about the incident?
? ?

45 Microwaving
The maximum cooking time that can be set on most microwaves is interesting:
1 hour, 40 minutes, 39 seconds
Can you explain why this is true?
? ?

46 Pricing Mystery

Suppose that you are a new employee at a candy store, and the owner has priced some of the confections as follows:

Ice Cream: 32 cents
Lollipop: 34 cents
Jawbreaker: 42 cents
Gum: 13 cents
Licorice: ?

The owner forgot to price the Licorice. Based upon the price of the other items, how much does the licorice cost?

? ?

47 An Old Riddle

What does man love more than life,
hate more than death or mortal strife?
It's that which contented men desire,
the poor possess, and the rich require,
the miser spends, the spendthrift saves,
and all men carry to their graves.

? ?

48 A Number Scheme

A	B	C	D
999	998	997	996
729	648	567	486
126	192	210	192
12	18	?	18

Using the top number in each vertical column A thru D as a beginning number, each succeeding number in the respective column is determined by the same technique.

What is the missing number indicated by the "?" ?

(The solution may require a "keen mathematical eye", but no difficult calculations.)

? ?

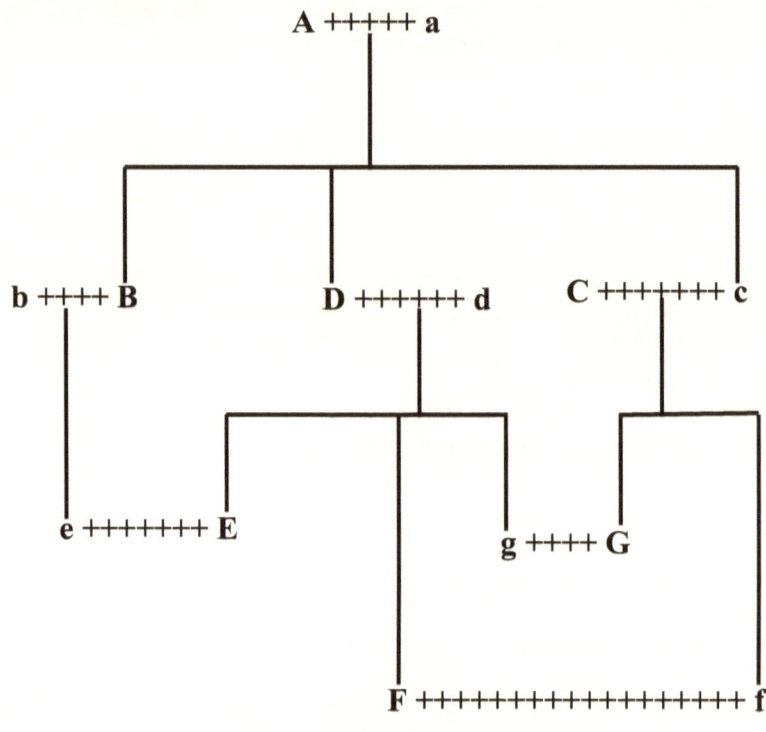

Key:
Capital letters represent husbands.
Lower-case letters represent wives.
"+++" represent marriages.

The Governor of Kgovjni (**F**) wants to give a small dinner party, inviting as his guests:
(1) His Father's Brother-in-law
(2) His Sister's Father-in-law
(3) His Father's Sister's Husband
(4) His Brother-in-law's Father
(5) His Grandfather's Son-in-law

The genealogy for these relationships are illustrated in the above diagram.

Using the letters of the diagram, identify the guests that are invited to the dinner party.
? ?

50 Geographic Extremes
What are the Northernmost, Southernmost, Easternmost, and Westernmost states of the United States by three separate categories?

(1) Name the four geographic extremes of the Contiguous 48 states.
(2) Name the four geographic extremes of the Continental 49 states.
(3) Name the four geographic extremes of the Total 50 states.
? ?

51 Spelling
In this puzzle, one, <u>and only one</u>, word is mispelled.
Can you identify the word?
PRIVILEGE
SKIING
CONSCIENTIOUS
BROCCOLI
VACUUM
SHERIFF
MEMORABILIA
CAFFEINE
BALLOON
SUPERSEDE
ACCOMMODATE
INSIDIOUS
HIPPOPOTAMUS
? ?

52 Two Answers
Determine a number that will logically complete the following sequence:
2,3,4,5,6,7,8,?
The number 9 is a valid choice. However, there is another number that also logically completes the sequence. What is that number?
? ?

53 Types of People
There are exactly (no more, no less) 10 types of people in this world:
(1) Those that understand Binary Math
(2) Those that do not
Can you explain this situation?
? ?

BLACK

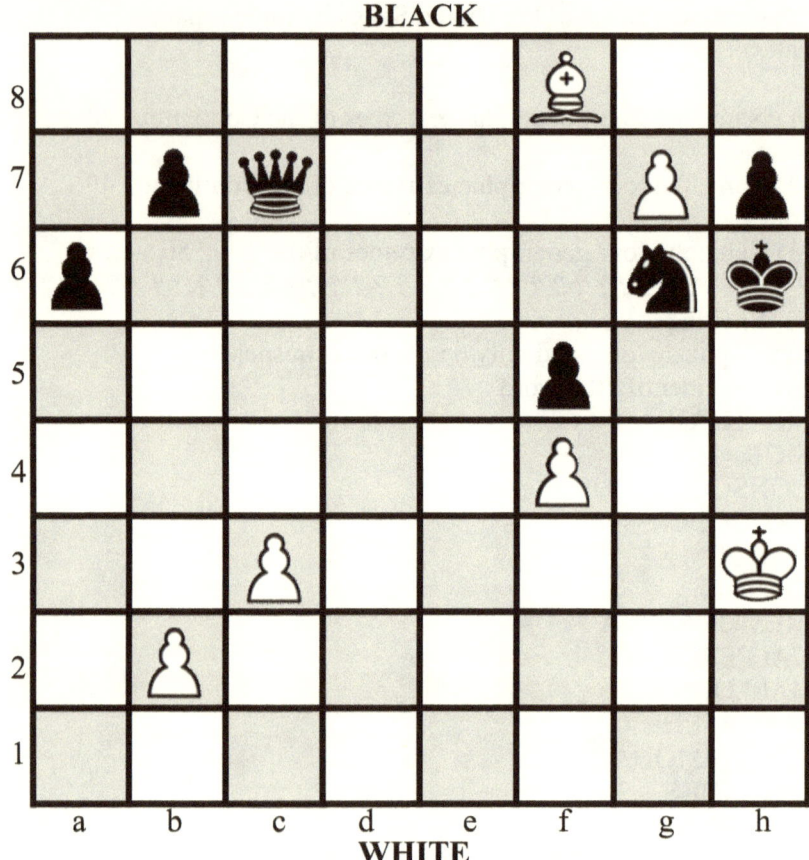

WHITE

(White's Move)
White declares a checkmate in two moves.
Is White correct?

? ?
55 The Greatest Riddle Of Them All
What is:
GREATER THAN JESUS;
MORE EVIL THAN SATAN;
POOR PEOPLE HAVE IT;
RICH PEOPLE WANT IT;
        and
IF YOU EAT IT, YOU WILL DIE?
(Hint: The answer is a seven-letter word.)
? ?

SECTION 3
EXPOSITIONS

There are *THREE* kinds of people:

Those who can do math, and those who cannot.

2
Thoughts on Infinity

Infinity is an interesting concept that implies qualities of endlessness, boundlessness, and limitlessness. It is a concept that represents the idea that some processes, like counting, can be continued without end. There is no smallest among the small, and no largest among the large, but always something still smaller, and something still larger.

∞

A graduate student at Trinity
Completed the square of infinity.
But it gave him the fidgets
To write down the digits,
So he dropped Math and took up Divinity.
                    Anonymous (author unknown)

∞

Fleas on dogs' backs are there to bite 'em,
But on those fleas are smaller fleas,
And on the smaller fleas are lesser fleas,
And so on ad infinitum.
                    Augustus DeMorgan
[from "Hoyt's New Cyclopedia of Practical Comments"
(1922), p.277.]

∞

There was a young lady named Bright
Whose speed was much faster than light.
She left home one day, in a relative way,
And returned the previous night.
                    Arthur Reginald Buller
[from the December 19, 1923 issue of "Punch".]

∞

There is a land of pure delight,
Where saints immortal reign;
Infinite day excludes the night,
And pleasures banish pain.
        Isaac Watts (1674-1748)
[from the first of six stanzas of a poetical hymn entitled
"A Prospect of Heaven makes Death Easy".]

3
Age Error Report

Dear Reader:

I received a letter from the Director of the United States Bureau of Vital Statistics located in Sweetgrass, Montana. He informed me that my birth date had been erroneously processed.

It seems that a gentleman named Hoo Ey (Chinese ancestry) made the mistake when he originally recorded the date of birth. The letter explained that Mr. Hoo Ey had a unique biological trait that likely contributed to the mistaken date. Hoo had <u>eight</u> fingers on each hand instead of the normal five. He would occasionally determine numbers in base 16 instead of base 10.

Therefore, according to the official DOB recording by the Bureau, my age is 50.
The Director assured me, however, that because $50_{16} = 80_{10}$, the original document recording would be corrected.

At least my Hooey Age is 50.

Merle
6/14/2015

4
Chess – A Plethora of Possibilities

The first move (white and black) in a game of chess can be performed in exactly 400 different ways. The possibilities for the first ten moves is more than 169 octillion, which is:
169,000,000,000,000,000,000,000,000,000.
A number followed by 27 zeros is really incomprehensible – so attempting to consider the total number of possible moves in any given lengthy chess game does overwhelm the imagination. But, back to my idea regarding the first ten moves that provide more than 169×10^{27} possibilities. How long would it take to play all possibilities for the first ten moves on each side? At the end of year 2013, the earth's population was approximately 7.2 billion people. On the basis of this population, and 169 octillion possibilities, more than 44 trillion years would be required to play all the possibilities, even if every man, woman, and child in the world played without cessation for that duration at the rate of one game per minute with no game repeated.

$$\frac{169 \times 10^{27} \, games}{\left(7.2 \times 10^9 \, people\right)\left(60 \, \frac{games}{hr}\right)\left(24 \, \frac{hrs}{day}\right)\left(365.25 \, \frac{days}{yr}\right)}$$

$$= 44{,}627{,}390{,}340{,}000 \, years$$

5
One Million Dollars

Suppose you are given the possibility of receiving One Million Dollars in One Dollar Bills. The <u>one</u> and <u>only</u> condition required for you to possess this million dollars is that you must personally, without any human or mechanical assistance (e.g., a suitcase on wheels), carry it away.

You have the opportunity to view the entire million dollars, and then determine how you will personally carry the money in a container (such as a suitcase or duffel bag) of your choice and specifications.

Will you be capable of possessing the money?

You will <u>not</u> be capable of possessing the money because you will not be capable of personally carrying the money. Why?

The reason is that each bill weighs about .035274 of an ounce. Therefore, one million dollars weighs $1,000,000 \times .035274 = 35,274$ ounces.
$35,274 \div 16 = 2,204.6$ pounds.
The one million dollars weighs more than a ton, or about 2,200 pounds.
This weight, of course, excludes the weight of the container.
No human is capable of carrying this weight. ☺

6
The Principles of a Relationship Between a Man and a Woman
(From a Mathematical Perspective)

Principle 1

To find a Woman, Time and Money are required.
Therefore,

Woman = Time x Money

Principle 2

"Time is Money." So,

Time = Money

Because time equals money, substituting money for time in Principle 1 gives us the
third principle:

Principle 3

A relationship with a Woman requires a lot of money.

Woman = (Money)^2

Principle 4

"Money is the root of all Problems." Therefore,

Money = $\sqrt{PROBLEMS}$

Principle 5

Squaring both sides of the equation in Principle 4 gives us the fifth principle.

$$(\text{Money})^2 = PROBLEMS$$

Using the equality of Principle 3 and substituting "Woman" in Principle 5 gives us
the logical and inerrant conclusion of Principle 6:

Principle 6

Woman = *PROBLEMS*

7
How to Split an Atom

The atom is tremendously minute – it's size is staggering and almost incomprehensible. Atoms are so small that it would take the entire population of the earth ten thousand years to count the number of them in one drop of water. Before one could be able to count them, one would have to shrink in size to one-billionth of an inch tall. An atom is so minuscule that about a hundred billion billion of them are contained on the head of a pin. The nucleus, or core of an atom, is about ten thousand times smaller than the atom, e.g., if the atom were expanded to the size of a concert hall, it's nucleus would be smaller than a housefly. The nuclei of all atoms are made up of protons and neutrons. These particles vary in number, depending on whether the nucleus belongs to an atom of gold, silver, etc.

To get past the electron barrier (minute particles that revolve about the atom), and break through into the nucleus of the atom, is the desired achievement of the "atom smasher". A successful attempt will result in a change in the number of protons and neutrons within the nucleus. Thus, by varying the number of protons and neutrons, new and different elements may be formed. This incredibly small grain of matter, i.e., the atom, is what the scientists predict to be the power source of the future.

Generally speaking, there are two methods by which an atom may be split. The first method, one very familiar to scientists, involves advanced technological processes. The splitting of the atom is accomplished by the use of accelerators, nuclear reactors, and cyclotrons. However, I would like to direct your attention to the second method – my method. My method is informal, and I prefer it to the technological method because it is much simpler.

Splitting the atom is a delicately difficult and extremely exacting process, of which there are three steps. The first step is securing the atom. The second step is the given specific conditions, and the last step is the actual splitting of the atom.

Isolating one atom is a fairly simple task, since there are trillions and trillions all around you. There is one important consideration in the selection of the atom. I will bring this to your attention in my next step. The two given conditions are that the atom must be frozen, and the important consideration is that you select a fairly large atom so that working with it will be a fairly easy task. The atom should be frozen because it will split much easier when it is hard. You may freeze the atom by placing it in the freezing compartment of your refrigerator. You will have less difficulty when observing these conditions, and the process will be fairly simple.

Having completed the two preceding steps, the atom is now ready to be actually split. For the completion of the process, two important and absolutely necessary tools required are a hammer and a fairly small chisel. The other important materials utilized are (1) a small table, (2) white paper, (3) tweezers, (4) minute file, (5) microscope, and (6) rubber gloves.

Select a small table on which to work. Cover the table with some plain white paper. This will aid in distinguishing the atom more readily. Using your tweezers, transfer the atom from the refrigerator, or wherever you may have placed it to be frozen. It is more efficient to use the tweezers when you are moving the atom, because it is somewhat awkward to use your fingers. Then, be viewing the atom with the microscope, it will be possible to cut a small groove on the top of the atom with the file. Rubber gloves may or may not be used in finishing the process. This can be left to the discretion of the individual administering this method. Then, with the assistance of the microscope, place the sharp end of the chisel in the filed groove of the atom. For the last and most important step, strike the chisel once, very sharply. Strike the chisel very sharply. This is extremely important. If the atom is not struck sharply and exactly, it will be necessary to strike the chisel many times in order to split the atom. This consistent pounding will cause the atom to become quite hot, because the movement of the particles within the atom's nucleus has increased in rapidity. As a result, upon splitting, the atom will release much more energy than normal. Correctly split, the atom will release energy amounting to no more than a few sparks. If one will pursue the correct method in this delicate process, the exact result will ensue.

As I shall now clarify, the foregoing method is not quite popular among the leading physicists and scientists of our country today. They and I are not in absolute accord as to the method used. However, this difference is of trivial consequence, and I shall spend no further timer speaking of it to you.

Research in Atomic Energy is providing valuable information. Scientists are learning about the nature and structure of atomic nuclei, and what forces operate within them, so as to acquire a fuller understanding of how atomic energy is released, and how this energy can be converted into work. Atoms contain gigantic power if their total energy could be transformed. For example, one pound of anything (uranium, coal, soil, etc.), would yield 11,400,000,000 kilowatt hours, or enough energy to provide the total electric requirements of the United States for several weeks. With the conversion and control of this elemental energy, future scientists may promote progress toward the development of atomic energy for constructive and peaceful purposes.

M.A.B.

8
Pythagorean Triples

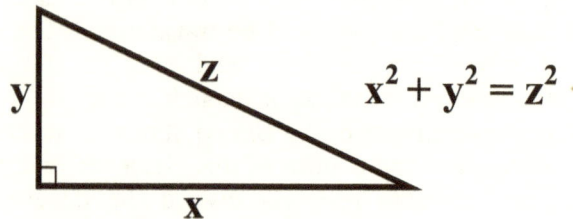

$$x^2 + y^2 = z^2$$

Given: x, y, and z are Positive Integers

Given: a and b are Positive Integers

Select a and b such that 2ab is an <u>Even</u> <u>Perfect</u> <u>Square</u>

Then, $x = a + \sqrt{2ab}$,

$\quad y = b + \sqrt{2ab}$, and

$\quad z = a + b + \sqrt{2ab}$

(x, y, z) is known as a Pythagorean Triple, and the numbers x, y, and z are listed in ascending sequence which indicates that "z" is the largest of the three numbers. Because it is the largest, it is <u>necessarily</u> the hypotenuse of the right triangle (x, y, z).

Example 1
Let 2ab = 100

Choose a = 5 and b = 10 so that (2)(5)(10) = 100

Therefore,

x = 5 + 10 = 15

y = 10 + 10 = 20

z = 5 + 10 + 10 = 25

The Pythagorean Triple is (15,20,25)

Example 2a
Let 2ab = 400

Choose a = 10 and b = 20 so that (2)(10)(20) = 400

Therefore,

x = 10 + 20 = 30

y = 20 + 20 = 40

z = 10 + 20 + 20 = 50

The Pythagorean Triple is (30,40,50)

Example 2b
Let 2ab = 400
Choose a = 4 and b = 50 so that (2)(4)(50) = 400
Therefore,
x = 4 + 20 = 24
y = 50 + 20 = 70
z = 4 + 50 + 20 = 74
The Pythagorean Triple is (24,70,74)

Example 3
Let 2ab = 17956 (134^2 =17956)
Choose a = 2 and b = 4489 so that (2)(2)(4489) = 17956
Therefore,
x = 2 + 134 = 136
y = 4489 + 134 = 4623
z = 2 + 4489 + 134 = 4625
The Pythagorean Triple is (136,4623,4625)

Note:
If 2ab is not an Even Perfect Square, then both a and b cannot be
integers, and the values of
x, y, and z will not all be Integers.

Example:
Let 2ab = 9
Then ab = 9/2
Choose a = ½ and b = 9
Therefore,
x = ½ + 3 = 3½
y = 9 + 3 = 12
z = ½ +9 +3 = 12½
The Pythagorean Triple is (3½,12, 12½)

Example of Pythagorean Triples

$$(3,4,5)$$
$$(5,12,13)$$
$$(6,8,10)$$
$$(7,24,25)$$
$$(8,15,17)$$
$$(9,12,15)$$
$$(9,40,41)$$
$$(10,24,26)$$
$$(14,48,50)$$
$$(15,20,25)$$
$$(24,70,74)$$
$$(30,40,50)$$
$$(136,4623,4625)$$

.
.
.

$$\infty$$

9
The Missing Dollar

Question:
Three guests check into a hotel room. The clerk says the bill is $30, so each guest pays $10. Later, the clerk realizes the bill should only be $25. To correct the situation, he gives the bellhop $5 to return to the guests. On the way to the room, the bellhop realizes that he cannot divide the money equally. Because the guests do not know the total of the revised bill, the bellhop decides to just give each guest $1 and keep $2 for himself. Each guest received a $1 refund. Now, each guest only paid $9, so the total paid is $27. The bellhop has $2, and $27 + $2 = $29. If the guests originally paid $30, what happened to the remaining $1?

$ $
Solution:
The initial payment of $30 is accounted for as the clerk receives $25, the bellhop takes $2, and the guests receive a $3 refund. The solution becomes easily apparent when the initial and net payments are written as simple equations.

Equation (1) defines what happened to the initial payment of $30:

(1) $30 (initial payment) = $25 (to clerk) + $2 (to bellhop) + $3 (refund)

Equation (2) defines the net payment after the refund of $3 is applied ($3 subtracted from both sides):

(2) $27 (net payment) = $25 (to clerk) + $2 (to bellhop)

Now, subtract $2 from both sides of equation (2):

(3) $25 (final payment) = $25 (to clerk)

The solution can be considered in the following manner:
The hotel receives $30 from the guests, and gives $5 to the bellhop to return to the guests. $30 minus $5 equals $25. $3 is given to the guests, and the bellhop keeps $2. Thus, $3 plus $2 equals $5, and the remaining $25 is in the possession of the hotel.

This is clearly not a paradox. Each guest has paid $9 for a total of $27. The $2 that the bellhop pilfered should have been subtracted (rather than added) for a total of $25 paid. So, $3 \times \$9 = \$27$, which accounts for the $25 room and the $2 taken by the bellhop.

There is no missing dollar.

10
Intersection Safety

Consider the concept of the relationships among the distance, rate, and time variables:

$$D = rt \text{, and}$$

$$t = \frac{D}{r}$$

Apply this concept to driving some vehicle through a very busy intersection.

Because the time varies inversely as the rate,

As $r \to \infty, t \to 0$, for any given D.

Therefore, the faster you go, the less time you spend in the intersection.

Conclusion:

If you go fast enough, you cannot possibly make contact with another vehicle – therefore, you will not be involved in an intersection accident.

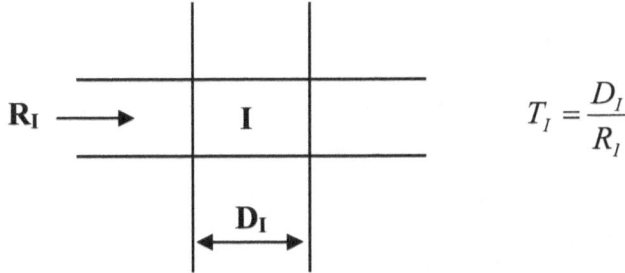

A logical way to avoid intersection accidents is to minimize travel time within the intersection.

In the diagram above,

     (1) I represents the Intersection.

     (2) D_I represents the Distance (in miles) across the intersection.

     (3) R_I represents the vehicle Rate of travel (in miles per hour).

     (4) T_I represents the Time (in hours) required to travel across the intersection.

Since $D_I = R_I \times T_I$, and D_I is a constant, the only way to decrease T_I is to increase R_I for a given distance D_I. Therefore, in order to achieve a minimum T_I for the intersection travel,

$$T_I(Min) = \lim_{R_I \to Max} \left(\frac{D_I}{R_I} \right)$$

In other words, in order to minimize T_I, it is necessary to maximize R_I.

For example, consider an intersection distance that is 50 feet (D_I =50 feet or .0095 mile). At an average rate of 10 miles per hour, between three and four seconds will be required to cross the intersection.

$$T_I = \frac{D_I}{R_I}$$

$$T_I = \frac{.0095 \, mile}{10 \, miles/hour} = .00095 \, hours$$

$T_I = .057$ minutes = 3.42 seconds

However, at a rate of 100 miles per hour, the time required to cross the intersection will be significantly less than the time above.

$$T_I = \frac{.0095 \, mile}{100 \, miles/hour} = .000095 \, hours$$

$T_I = .0057$ minutes = .342 seconds

Increasing the rate by a factor of ten has reduced the intersection time by a factor of ten. Since the time spent in the intersection is ten times less, it is evident that the probability of being involved in an accident is also ten times smaller. Further increasing the rate to some practical maximum value will obviously assure an even safer journey across the intersection!

HAPPY MOTORING!!

11
Heliocentric Parallax

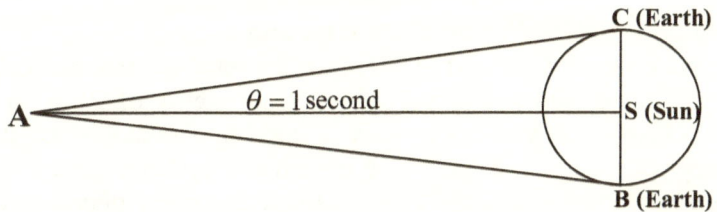

For the very closest stars, we can use a type of trigonometry known as parallax. This concept uses the apparent change in a star's position caused by the Earth moving from one endpoint of the diameter of the orbit around the Sun to the other endpoint of that diameter in order to triangulate the star's positional distance.

$\theta = 1\, \text{second}$

$CS \approx 93 \times 10^6\, mi$

$$\tan\theta = \frac{CS}{AS}$$

$$\tan 1\, \text{second} \approx \frac{93 \times 10^6\, mi}{AS\, mi}$$

$$(\tan 1\, \text{second})(AS\, mi) \approx 93 \times 10^6\, mi$$

$$\left(4.85 \times 10^{-6}\right)(AS\, mi) \approx 93 \times 10^6\, mi$$

$$AS\, mi \approx \frac{93 \times 10^6\, mi}{4.85 \times 10^{-6}}$$

$$AS\, mi \approx 1.918 \times 10^{13}\, mi \approx 1\, \text{parsec}$$

$$\frac{1.918 \times 10^{13}\, mi}{(186{,}300\, mi/\text{sec})(86{,}400\, \text{sec}/day)(365.25\, day/yr)} \approx 3.26$$

light-years

The Parsec is a unit of distance of interstellar space equal to a heliocentric parallax of one second of arc, equivalent to 206,265 times the distance from the earth to the sun, or 3.26 light-years.

12
Perspectives of Magnitudes

The total U.S. Government Debt at the end of 2013 was about 17 trillion dollars. At the end of FY2014, the total government debt in the United States is expected to be in excess of 21 trillion dollars. Now, do you really have a good understanding of the concept of a trillion dollars, or a trillion of anything?; or even a billion?; or even a million? I do not think that many people have a good understanding of the relative magnitudes of very large or very small numbers.

Let's put these numbers in perspective. First, a TRILLION from a perspective of time:
If you counted by 1s from 1 to a trillion at the rate of one number each second, what time would be required? Sure, you could count faster than one number per second, but I am thinking about an <u>average</u> of one number per second, because I will require you to count unceasingly every second of every hour of every day without stopping to eat or sleep, or anything. Now, under those conditions, how long would it take you to get to a trillion? I will document the answer:

How many seconds in a day?

$$(60\,sec/min)(60\,min/hr)(24\,hr/day) = 86,400\,sec/day$$

Since we have one trillion seconds to count, we divide one trillion seconds by the number of seconds in a day to determine how many days necessary to count to a trillion.

$$\frac{1,000,000,000,000\ sec}{86,400\,sec/day} = 11,574,074.07\ days$$

Convert days to years:

$$(1)\quad \frac{11,574,074.07\ days}{365.25\ days/yr} = 31,688.0878\ yr$$

Obviously, you do not have more than 31,000 years to devote to counting to a trillion.
Let's look at a trillion in a different way – from the perspective of distance:
The length of a dollar bill is 6.2 inches. Imagine aligning one trillion dollar bills end-to-end. What distance would be required

for these trillion dollar bills? One trillion can be expressed as 10^{12}.

$$\left(10^{12}\right)\!\left(6.2\ in\right)=6.2\times10^{12}\ in$$

Now, how far is 6,200,000,000,000 inches?
Convert inches to miles:
$$\left(6.2\times10^{12}\ in\right)\!\left(1\ ft/12\ in\right)\!\left(1\ mi/5280\ ft\right)=97,853,535.35\ mi$$
Assume that the earth is a perfect sphere with a circumference of 25,000 miles at the equator. When we divide the distance of the one trillion dollar bills by the distance around the equator, the result is the number of times the dollar bills will extend around the earth.

$$(2)\ \ \frac{97,853,535.35\ mi}{25,000\ mi}=3,914.1414\ times$$

Yes, those dollar bills will circle the earth 3,914 times plus almost 3600 miles.

So much for a trillion. Let's examine the same situations from the perspective of a BILLION:
A billion is 1000 times smaller than a trillion. Therefore, if answers (1) and (2) of the trillion section are divided by 1000, the results are correct answers for a billion.

(1) The time required to count to a billion becomes:
$$\frac{31,688.0878\ yr}{1000}=31.6880878\ yr$$
I will assume that although you realistically may have almost 32 years available, you will not want to devote it to this task.

(2) The distance required by one billion dollar bills end-to-end:
$$\frac{3,914.1414\ times}{1000}=3.9141414\ times$$
So, a billion dollar bills will extend almost 4 times around the world.

Again, let's examine the same situations from the perspective of a MILLION:
A million is 1000 times smaller than a billion. Therefore, if answers (1) and (2) of the billion section are divided by 1000, the results ar correct answers for a million.

(1) The time required to count to a million becomes:
$$\frac{31.6880878\,yr}{1000} = .0316880878\,yr$$
Convert this fraction of a year to days:
$$(365.25\,days/yr)(.0316880878\,yr) = 11.57407\,days$$
You could certainly devote between 11 and 12 days to the task if you are so inclined.

(2) The distance required by one million dollar bills end-to-end:
$$\frac{(3.9141414\,times)(25,000\,mi)}{1000} = 97.853535\,mi$$
A million dollar bills will extend almost 100 miles – merely a pleasant two hour journey if you average 50 mph.

Summary:

Perspective		
Magnitude	Time (Count)	Distance (Dollar Bills)
Trillion 10^{12}	31,688 yr	3,914 times around the earth
Billion 10^{9}	31.688 yr	3.914 times around the earth
Million 10^{6}	11.57 days	97.85 mi

13
The Divine Proportion

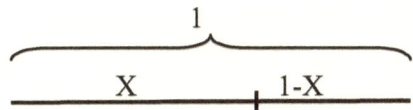

$$\frac{1}{X} = \frac{X}{1-X}$$

X is the mean proportional, or geometric mean, between 1 and 1-X.

$$X^2 = 1 - X$$

$$X^2 + X - 1 = 0$$

Solving for X using the quadratic formula yields two values for X:

$$X = \frac{-1 \pm \sqrt{1+4}}{2}$$

(1) $X = \dfrac{-1+\sqrt{5}}{2} = .6180339887$

(2) $X = \dfrac{-1-\sqrt{5}}{2} = -1.6180330887$

Because the value of X in (2) is negative, it is disregarded.

The reciprocal of (1) is $X^{-1} = \dfrac{1}{X}$

$$\frac{1}{X} = \left(\frac{-1+\sqrt{5}}{2}\right)^{-1} = \frac{1}{\dfrac{-1+\sqrt{5}}{2}} = \frac{2}{\sqrt{5}-1} = 1.6180339887$$

$\dfrac{2}{\sqrt{5}-1}$ can be shown to be equivalent to $\dfrac{\sqrt{5}+1}{2}$ by rationalizing the denominator:

$$\left(\frac{2}{\sqrt{5}-1}\right)\left(\frac{\sqrt{5}+1}{\sqrt{5}+1}\right) = \frac{2\sqrt{5}+2}{5-\sqrt{5}+\sqrt{5}-1}$$

$$= \frac{2\left(\sqrt{5}+1\right)}{4} = \frac{\sqrt{5}+1}{2} = 1.6180339887$$

This value is known as the Golden Ratio, or
Golden Mean, or
Golden Proportion, or
Divine Proportion.

Applying the Golden Ratio concept to a rectangle, the Golden Rectangle can be established.

$$\frac{l}{w} = \frac{l+w}{l}$$

$$l^2 = wl + w^2$$
$$l^2 - wl - w^2 = 0$$

$$l = \frac{w \pm \sqrt{w^2 + 4w^2}}{2}$$

$$(3)\ l = \frac{w \pm w\sqrt{5}}{2}$$

Solve (3) for w:

$$l = \frac{w\left(1 \pm \sqrt{5}\right)}{2}$$

$$2l = w\left(1 \pm \sqrt{5}\right)$$

$$(4)\ w = \frac{2l}{1 \pm \sqrt{5}}$$

If you specify a particular width (w) for a rectangle, the length (l) is determined by using the positive root of equation (3).
Conversely, if you specify a particular length (l) for a rectangle, the width (w) is determined by using the positive root of equation (4).

In God's creation, there exists a "Divine Proportion" that is manifested in a multitude of shapes, numbers, and patterns whose relationship can only be the result of the omnipotent, good, and all-wise God of Scripture. This Divine proportion – existing in the smallest to largest entities, in living and also in non-living things – reveals the awesome handiwork of God and His interest in beauty, function, and order. In addition to its occurrences in nature, art, and architecture, the golden rectangle is used in advertising and merchandising to appeal to people's aesthetic values of size and proportion.

14
What Part of a Car Travels Farthest?

Reference the Earth Diagram below for the following explanation:

In determining the answer to this question, consider some specified period of time or distance (e.g., 2 hours, 24 hours, 2 weeks, 100 miles, 1,000 miles, 10,000 miles, etc.), or even the period of time equal to the life of a given vehicle.

One suggested answer to this question is the FRONT BUMPER. This is false. For any specified traveled distance, both bumpers will travel the same distance if they are both the same distance from the ground. If both bumpers are the same distance above the ground, they will travel <u>exactly</u> the same distance. However, if the rear bumper is higher than the front bumper, the rear bumper will travel farther than the front bumper for any specified distance. Why? Reference the next paragraph.

As the car travels in any direction along the surface of the earth, a point on the TOP OF THE CAR, and a point on the car's bumper will trace arcs of two different circles. The largest circle represents the path of a point on top of the car The smaller circle represents the path of a point on the car bumper. T is the radius of the largest circle; B is the radius of the smaller circle. Assuming that the top of the car is 4 feet above the bumper, a point on the top of the car will travel about .001 of a foot farther than a point on the bumper in a distance of one mile. This is a small difference, but it <u>is</u> a <u>greater</u> distance.

Earth Diagram

$T > B$

$C = 2\pi r$

$\therefore 2\pi T > 2\pi B$

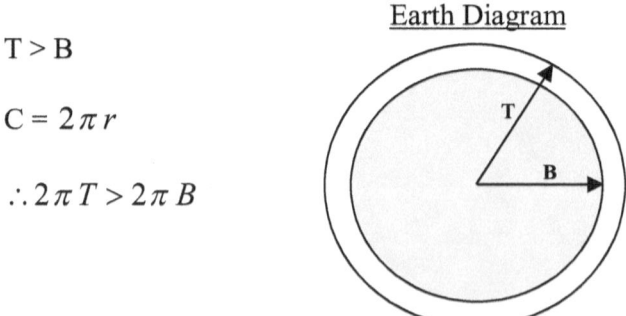

The point on the top of the car travels farther than any other point of the car.

It has been suggested that the WHEELS travel the farthest. This is not true. For example, consider a car with 20-inch wheels. A specific point on the circumference of the tire will travel approximately 5.24 feet in one revolution. That specific point, in traveling one mile, will make about 1008 revolutions, but will travel only one mile. Many other points of the car above the tires will travel farther, especially the top of the car.

Another answer submitted for this question was the car KEYS. This is incorrect reasoning because the car keys are not a physical, permanent part of the car. But, just for fun, consider the car keys as part of the car. There are many parts of the car that travel farther than the keys during the time that the keys are in the ignition. When the car is parked and the keys are placed in a pocket or purse, the keys are traveling some distance farther than the car. However, the radius of the arc traced by the keys will be less than the radius of the arc traced by the top of the car. Consequently, the additional distance traversed by the keys in a pocket or purse generally will not exceed the greater distance traversed by the car top, unless the person possessing the keys now continues to travel as a passenger in another vehicle on a relatively LONG journey. The distance traveled by the keys under these circumstances could easily exceed the traveled distance of the car top. This would be true especially if the journey involved flying in a jetliner several miles above the earth's surface. In this case, the radius of the arc traced by the keys is considerably greater, and therefore the distance traveled by the keys easily would be greater than the distance that was traveled by the car top.

P.S.: Consider this – In a lifetime of walking and running, your head travels farther than your feet. (Could this be the source of our headaches?) ☹

15
PERFECT BRIDGE HANDS

(1) What are the chances of all four players receiving perfect hands in a bridge game?

(2) What are the odds against such an occurrence?

Let A = the probability of the first player receiving a perfect hand.

Let B = the probability of the second player also receiving a perfect hand.

Let C = the probability of the third player also receiving a perfect hand.

Let D = the probability of the fourth player also receiving a perfect hand.

$P(A) = 4 \div {}_{52}C_{13}$

$P(B) = 3 \div {}_{39}C_{13}$

$P(C) = 2 \div {}_{26}C_{13}$

$P(D) = 1 \div {}_{13}C_{13}$

${}_{52}C_{13} = 52! \div (39! \, 13!) = 635,013,559,600$

${}_{39}C_{13} = 39! \div (26! \, 13!) = 8,122,225,444$

${}_{26}C_{13} = 26! \div (13! \, 13!) = 10,400,600$

${}_{13}C_{13} = 13! \div (0! \, 13!) = 1$

The answer to (1) is:

$$P(A \text{ and } B \text{ and } C \text{ and } D) = [4 \div {}_{52}C_{13}] \times [3 \div {}_{39}C_{13}]$$

$$\times [2 \div {}_{26}C_{13}] \times [1 \div {}_{13}C_{13}]$$

$= 24 \div 53{,}644{,}737{,}765{,}488{,}792{,}839{,}237{,}440{,}000$
$= 1 \div 2{,}235{,}197{,}406{,}895{,}366{,}368{,}301{,}560{,}000$

Now consider question (2):

Let h = the number of ways an event occurs successfully.

Let f = the number of ways an event occurs unsuccessfully.

Let p = the probability of success.

Let q = the probability of failure.

$p = h \div (h + f)$

$q = f \div (h + f)$

The odds in favor of an event is defined as $p \div q = h \div f$

The odds against an event is defined as $q \div p = f \div h$

The answer to (2) is:

The odds against the occurrence of four perfect bridge hands is:

$f \div h = 53{,}644{,}737{,}765{,}488{,}792{,}839{,}237{,}439{,}976 \div 24$
$\qquad = 2{,}235{,}197{,}406{,}895{,}366{,}368{,}301{,}559{,}999$ to 1

The estimated population of the world at the end of 2013 was 7.2 billion.

$7,200,000,000 \div 4 \approx 1,800,000,000$ (number of bridge foursomes).

$1,800,000,000 \times 120$ hands/day $\approx 216,000,000,000$ hands/day.

This is the number of bridge hands (foursomes) dealt at the rate of one hand every 12 minutes, or 120 hands per 24-hour day.

Because of the fact that there is only 1 chance out of every 2,235,197,406,895,366,368,301,560.000 hands for a perfect bridge hand to occur,

2,235,197,406,895,366,368,301,560.000 hands $\div 216,000,000,000 \approx 1.034813614 \times 10^{16}$ days that would be required before the expectation of the occurrence of a perfect bridge hand.

$(1.034813614 \times 10^{16}$ days$) \div (365$ days/year$)$
$\approx 2.835105792 \times 10^{13}$ years.

$2.835105792 \times 10^{13}$ years $\approx$ 28 + trillion years before a perfect bridge hand (foursome) could be expected to occur – even though the entire population of the world is playing 24 hours per day.

16
Energy Plus

Assume the technological capability exists to completely disintegrate a given mass of <u>any</u> matter in order to use the liberated energy for peaceful applications (this necessarily means a controlled process). The energy release from the total disintegration (annihilation) of a given mass is explained by Einstein's mass-energy relationship,

$$E = mc^2$$

The U.S. Customary System of units will be used in the following example, where

E (energy) is in ft • lbs

> The energy required to move one pound a distance of one foot is defined as a foot • pound of energy.

m (mass) is in slugs (1 slug $= \dfrac{1\,lb}{ft/\sec^2}$)

> Because of the fact that w = mg, it is convenient to work problems in this system by writing the mass as $\dfrac{w}{g}$,
>
> where w is the weight, and g is the acceleration of gravity.

c is the speed of light ($9.836 \times 10^8 \dfrac{ft}{\sec}$)

Suppose that a one pound mass of some matter is completely converted to energy. How many foot • pounds of energy are generated?

$$E = mc^2$$

$$E = \left(\frac{1\,lb}{32.174\,ft/\sec^2} \right) \left(9.836 \times 10^8 \frac{ft}{\sec} \right)^2$$

$$E = \left(3.1 \times 10^{-2} \frac{lb\,\sec^2}{ft} \right) \left(9.7 \times 10^{17} \frac{ft^2}{\sec^2} \right)$$

$$E = 3.0 \times 10^{16}\,ft \bullet lbs$$

$$E = 30{,}000{,}000{,}000{,}000{,}000\,ft \bullet lbs$$

This amount of energy will move $3 \times 10^{16}\, lbs$ a distance of one foot, or one pound a distance of

$3 \times 10^{16}\, ft$, or any product combination that equals $3 \times 10^{16}\, ft \bullet lbs$.

To better understand the magnitude of this amount of energy, consider a large aircraft carrier. The U.S. Nimitz-class carriers are among the largest warships in the world. The several ships in this category are nuclear-powered and each carrier weighs more than 100,000 long tons $(2240\, lbs/ton)$. The energy from one pound of mass would propel one of these ships 1.02 times around the world at the equator (that's about 25,500 miles).

The calculations follow:
Assume the circumference of the earth at the equator is 25,000 miles.
$$\left(25 \times 10^3\, mi\right)\left(5280\, ft/mi\right) = 1.32 \times 10^8\, ft$$

$$\frac{\left(3.0 \times 10^{16}\, ft \bullet lbs\right)}{\left(100 \times 10^3\, tons\right)\left(2.24 \times 10^3\, lbs/ton\right)} = 1.34 \times 10^8\, ft$$

$$\frac{1.34 \times 10^8\, ft}{1.32 \times 10^8\, ft} = 1.02\ times$$

Another illustration will determine how much water this same amount of energy can lift to a height of one mile. The calculations below indicate that the energy from a one pound mass can lift 5.7×10^{12} pounds of water one mile. 5.7×10^{12} pounds of water is a volume equivalent to a cube of water whose edge is approximately 4,500 ft (that's a cube of water whose edge is about 85% of a mile).

$$\frac{3.0 \times 10^{16}\, ft\, lbs}{5280\, ft} = 5.7 \times 10^{12}\, lbs = 5,700,000,000,000\, lbs$$

1 pound of water $= 1.6020 \times 10^{-2}\, ft^3$, or $1 ft^3 = 62.4\, lbs$
$$\left(5.7 \times 10^{12}\, lbs\right)\left(1.6 \times 10^{-2}\, ft^3/lb\right) = 9.1 \times 10^{10}\, ft^3 = 91,000,000,000\, ft^3$$

$$\sqrt[3]{9.1 \times 10^{10}\, ft^3} = 4,498\, ft = \text{edge of cube}$$

Imagine the small quantities of anything (dirt, water, vegetation, garbage, etc.) that could be used to provide all of our energy needs. There would never be a concern for the shortage of any of our resources. This would mean inexhaustible supplies of matter for all the energy we could need or want. The accomplishment of this process to completely convert mass to energy is possible within the reality of the physical laws and relationships designed by the Living Lord God for His universe. The realization of this process is questionable, however, because of our intellectual and technological limitations.

Notes

☞ F (Force) = m (mass) × a (acceleration)

☞ In the U.S. Customary System, the unit of force is given as 1 lb, the unit of acceleration is given as $1 ft/\sec^2$, and the unit of mass is then appropriately defined as the mass of a body whose acceleration is $1 ft/\sec^2$ when the resultant force on the body is 1 lb. This unit of mass is called one slug. (Perhaps the origin of the word was the concept of mass as inertia or sluggishness).

☞ F (pounds) = m (slugs) × a $\left(ft/\sec^2\right)$

☞ Work = Force × Distance

☞ The foot ● pound is a measure of energy or work

17
Analog Time Intervals
Hour and Minute Hand Interceptions

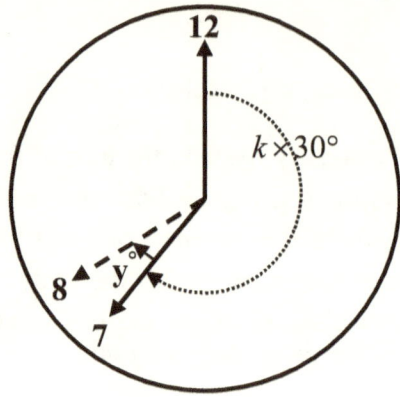

Let R = Rate of Minute Hand = $360°/hr \ or \ 6°/min$.

r = Rate of Hour Hand = $30°/hr \ or .5°/min$.

k is the hour variable.

$k \times 30°$ = the angular distance between the two hands.

$y°$ = the angular distance traveled by the hour hand.

At the moment that the minute hand overtakes the hour hand, they both will have traveled for the same length of time.

$$D_R = k \times 30° + y° \ ; D_r = y°$$

$$t_R = \frac{D_R}{R_R} \ ; t_r = \frac{D_r}{r_r}$$

Since $t_R = t_r$, $\dfrac{k \times 30° + y°}{6°/min} = \dfrac{y°}{.5°/min}$

$$\frac{(6y)^{°2}}{min} = \frac{k15^{°2} + .5y^{°2}}{min}$$

$$(6y)^{°2} = k15^{°2} + .5y^{°2}$$

$$(6y)^{°} = k15° + .5y°$$

$$5.5y° = k15° \quad y° = \frac{k15°}{5.5}$$

$(k \times 30° + y°) =$ the angle of interception

$5 \min = 30°$

$5/30 \min = 1/6 \min = 1°\,;\, 1 \min = 6°$

$$(k \times 30° + y°)° \times \frac{1 \min}{6°} = \frac{k \times 30° + y°}{6} \min$$

So, $\dfrac{(k \times 30° + y°)}{6} =$ minute of interception

There are eleven (11) interceptions forming eleven central angles. Each interception is exactly 65.454545... minutes from the two adjacent interceptions.

$\dfrac{360°}{11} = 32°.727272... =$ the size of each central angle.

Since $1° = 2 \min$ (hour hand),

$32°.727272... = 2 \times 32°.727272... = 65.454545$ minutes

So, $\dfrac{360°}{11} \times \dfrac{2 \min}{1°} = \dfrac{720}{11}$ minutes $= 65.454545...$ minutes $=$

distance between two interceptions

Interception Intervals

k	I_m	I_h
1	65.4545...	1.0909...
2	130.9090...	2.1818...
3	196.3636...	3.2727...
4	261.8181...	4.3636...
5	327.2727...	5.4545...
6	392.7272...	6.5454...
7	458.1818...	7.6363...
8	523.6363...	8.7272...
9	589.0909...	9.8181...
10	654.5454...	10.9090...
11	720	12

k = the hour variable preceding the interception (there are 11 interceptions)

I_m = interception time in <u>minutes</u>

I_h = interception tune in <u>hours</u>

$$I_m = \left(\frac{720}{11}\right)(k)$$

$$k \in \left\{k \mid k \in J^+ \wedge 1 \le k \le 11\right\}$$

$$I_h = \frac{\dfrac{720}{11}(k)}{60}$$

18
INSUFFICIENT FORCE

"And so, no force, however great,
can stretch a cord, however fine,
into a horizontal line
that shall be absolutely straight."
William Whewell
(1794-1866)

This expression is found in the "Elementary Treatise on Mechanics" (1819) in a presentation regarding the equilibrium of forces on a point. It is reputed to be an example of unconscious, but perfect rhyme. An illustration of the physics involved in this concept is as follows:

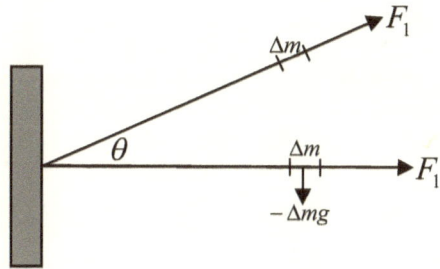

$$F_R = F_1 Sin\theta - \Delta mg$$

$$F_R = \lim_{\theta \to 0}(F_1 Sin\theta - \Delta mg) = -\Delta mg$$

Consider a cable attached securely at one end. A force F_1, directed away from the secure end, is applied at the opposite end of the cable. The magnitude of the force and the angle θ of the force from the horizontal (horizontal with respect to the point of attachment) can be such that any small mass Δm of the cable is above the horizontal. As θ approaches 0, i.e., as the line of force approaches the horizontal, the resultant force on Δm will vary continually. When θ is 0, i.e., when the force is horizontal, the resultant force on Δm is negative. Consequently, the mass Δm is below the horizontal.

To summarize this concept:

There is <u>no</u> horizontal force (even an infinitely great force) that can prevent a horizontal line (cord, cable, etc.) from being pulled below the horizontal. In other words, because the force has no upward vertical component, the force of gravity is affecting the line uncontested, and therefore the line will sag below the horizontal.

19
Merlen Formula

The Merlen formula was developed to determine the validity of the six variables of <u>any</u> triangle. The general formula relates all six variables of a triangle by equating an expression of the three sides with an expression of the three angles. By substituting the values of the sides and angles into the general formula (6), the accuracy of the variables can be verified. If there is at least one incorrect variable, note that the General Formula will not indicate which particular variables are incorrect, but only that the values of the six variables are inconsistent.

Equations 7,8,9,11,12, and 13 following the General Formula were derived to calculate the value of each specific variable as it relates to the remaining five variables. By applying the appropriate formula, the correct value of any variable can be calculated relative to the other five variables. Reference the Appendix for information regarding calculations.

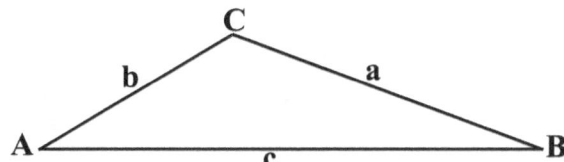

(1) $a^2 = b^2 + c^2 - 2bc \cos A$

(2) $a^2 = b^2 + c^2 - 2\left(\dfrac{a \sin B}{\sin A}\right)\left(\dfrac{a \sin C}{\sin A}\right)\cos A$

(3) $a^2 = b^2 + c^2 - \dfrac{2a^2 \sin B \sin C \cos A}{\sin^2 A}$

(4) $a^2 + \dfrac{2a^2 \sin B \sin C \cos A}{\sin^2 A} = b^2 + c^2$

(5) $a^2\left(1 + \dfrac{2 \sin B \sin C \cos A}{\sin^2 A}\right) = b^2 + c^2$

(6) $\dfrac{b^2 + c^2}{a^2} = 1 + \dfrac{2\sin B \sin C \cos A}{\sin^2 A}$ [General Formula]

From (5), Solve for a:

$$a^2 = \dfrac{b^2 + c^2}{1 + \dfrac{2\sin B \sin C \cos A}{\sin^2 A}}$$

$$a^2 = \dfrac{b^2 + c^2}{\dfrac{\sin^2 A + 2\sin B \sin C \cos A}{\sin^2 A}}$$

$$a^2 = \dfrac{\left(b^2 + c^2\right)\left(\sin^2 A\right)}{\sin^2 A + 2\sin B \sin C \cos A}$$

(7) $a = \sin A \sqrt{\dfrac{b^2 + c^2}{\sin^2 A + 2\sin B \sin C \cos A}}$

From (4), Solve for b:

$$b^2 = a^2 - c^2 + \dfrac{2a^2 \sin B \sin C \cos A}{\sin^2 A}$$

$$b = \sqrt{a^2 - c^2 + \dfrac{2a^2 \sin B \sin C \cos A}{\sin^2 A}}$$

$$b = \sqrt{\dfrac{\sin^2 A\left(a^2 - c^2\right) + 2a^2 \sin B \sin C \cos A}{\sin^2 A}}$$

(8) $b = \dfrac{1}{\sin A}\sqrt{\sin^2 A\left(a^2 - c^2\right) + 2a^2 \sin B \sin C \cos A}$

From (4), Solve for c:

$$c^2 = a^2 - b^2 + \frac{2a^2 \sin B \sin C \cos A}{\sin^2 A}$$

$$c = \sqrt{\frac{\sin^2 A(a^2 - b^2) + 2a^2 \sin B \sin C \cos A}{\sin^2 A}}$$

(9) $c = \dfrac{1}{\sin A}\sqrt{\sin^2 A(a^2 - b^2) + 2a^2 \sin B \sin C \cos A}$

From (3), Solve for A:

$$a^2 = b^2 + c^2 - \frac{2a^2 \sin B \sin C \cos A}{\sin^2 A}$$

$$\frac{2a^2 \sin B \sin C \cos A}{\sin^2 A} = b^2 + c^2 - a^2$$

Substitute value of Cos A from (1):

$$\frac{2a^2 \sin B \sin C}{\sin^2 A}\left(\frac{b^2 + c^2 - a^2}{2bc}\right) = b^2 + c^2 - a^2$$

$$\frac{2a^2 \sin B \sin C}{\sin^2 A} = (b^2 + c^2 - a^2)\left(\frac{2bc}{b^2 + c^2 - a^2}\right)$$

(10) $\sin^2 A = \dfrac{2a^2 \sin B \sin C}{2bc}$

$$\sin A = a\sqrt{\frac{\sin B \sin C}{bc}}$$

(11) $A = \sin^{-1}\left[a\sqrt{\dfrac{\sin B \sin C}{bc}}\right]$

From (10), Solve for B:

$$\sin^2 A = \frac{2a^2 \sin B \sin C}{2bc}$$

$$2bc \sin^2 A = 2a^2 \sin B \sin C$$

$$\sin B = \frac{2bc \sin^2 A}{2a^2 \sin C}$$

$$\sin B = \frac{bc \sin^2 A}{a^2 \sin C}$$

$$(12) \quad B = \sin^{-1}\left[\frac{bc \sin^2 A}{a^2 \sin C}\right]$$

From (10), Solve for C:

$$\sin^2 A = \frac{2a^2 \sin B \sin C}{2bc}$$

$$2bc \sin^2 A = 2a^2 \sin B \sin C$$

$$\sin C = \frac{2bc \sin^2 A}{2a^2 \sin B}$$

$$\sin C = \frac{bc \sin^2 A}{a^2 \sin B}$$

$$(13) \quad C = \sin^{-1}\left[\frac{bc \sin^2 A}{a^2 \sin B}\right]$$

Determine the Validity of the Six Variables of a Triangle

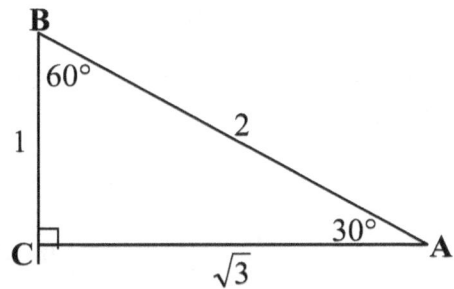

Use Formula (6) to verify the six variables of the above triangle:

$$\frac{b^2 + c^2}{a^2} = 1 + \frac{2\sin B \sin C \cos A}{\sin^2 A}$$

Does $\dfrac{\left(\sqrt{3}\right)^2 + 2^2}{1} = 1 + \dfrac{(2)(.8660254038)(1)(.8660254038)}{.25}$?

$$7 = 1 + \frac{(2)(.8660254038)^2}{.25}$$

$$7 = 1 + \frac{1.5}{.25}$$

$$7 = 1 + 6$$

$$7 = 7$$

Appendix

Relationships Between Degree of Accuracy of Calculated Lengths and Angles

When solving a triangle for any of its variables, the following should be observed:

Length to:	Angle to:
2 Significant Digits	Nearest 30 Minutes $= 0.5°$
3 Significant Digits	Nearest 5 Minutes $= 0.083°$
4 Significant Digits	Nearest 1 Minute $= 0.0167°$
5 Significant Digits	Nearest 0.1 Minute $= 0.00167°$

20
WORD TAXONOMY

A Taxonomy is the science or technique of classification. The concept of Taxonomy is usually applied to the study of plants and animals. I have developed a word taxonomy of the English Language. During my many years of life and education, I have not seen, in any text (high school or university), any documented form of a Word Taxonomy.

Now, I will briefly digress to explain, from a mathematical perspective, the theory of my development. When comparing any two words with respect to three different elements (Spelling, Pronunciation, Meaning), there are exactly eight possible combinations of classification [$2^3 = 8$].

In the Word Taxonomy table illustrated on the following page, the three elements of classification are in the extreme left column, the eight classifications are listed in the top row, and examples of each classification are given in the bottom row of the table.

WORD TAXONOMY

CLASSIFICATION → / ELEMENTS	HOMONYM	HOMOPHONE	HOMOGRAPH	SYNONYM	SAME WORD	DIFFERENT WORD*	PRONUNCIATION VARIATION	SPELLING VARIATION
SPELLING	=	≠	=	≠	=	≠	=	≠
PRONUNCIATION	=	=	≠	≠	=	≠	≠	=
MEANING	≠	≠	≠	=	=	≠	=	=
EXAMPLES (ELEMENTS →)	RAIL, QUARRY	HAIR/HARE, TOO/TWO	POLISH, CONDUCT	BRIEF/TERSE, AVOID/ESCHEW	EQUAL	WATER/HOUSE, COLD/HOT	CREEK, LABORATORY	GAGE/GAUGE, OCHER/OCHRE

SPELLING, PRONUNCIATION, AND MEANING ARE ELEMENTS THAT CLASSIFY
VARIOUS CATEGORIES OF WORDS

Homonyms are words that have the same spellings, the same pronunciations, and different meanings (e.g., Rail).

Homophones are words that have different spellings, the same pronunciations, and different meanings (e.g., Hair/Hare).

Homographs are words that have the same spellings, different pronunciations, and different meanings (e.g., Polish).

Synonyms are words that have different spellings, different pronunciations, and the same meanings (e.g., Brief/Terse).

Same Words are words that have the same spellings, the same pronunciations, and the same meanings (e.g., Equal).

Different Words* are words that have different spellings, different pronunciations, and different meanings (e.g., Water/House).

Pronunciation Variations are words that have the same spellings, different pronunciations, and the same meanings (e.g., Creek).

Spelling Variations are words that have different spellings, the same pronunciations, and the same meanings (e.g., Gage/Gauge).

* Note: *Antonyms* (words with opposite meanings) are included in this category.

Merle

95

APPENDIX 1
ANSWERS FOR
MATHEMATICS – SECTION 1

1
The Bookworm
An Arithmetic & Logic Problem

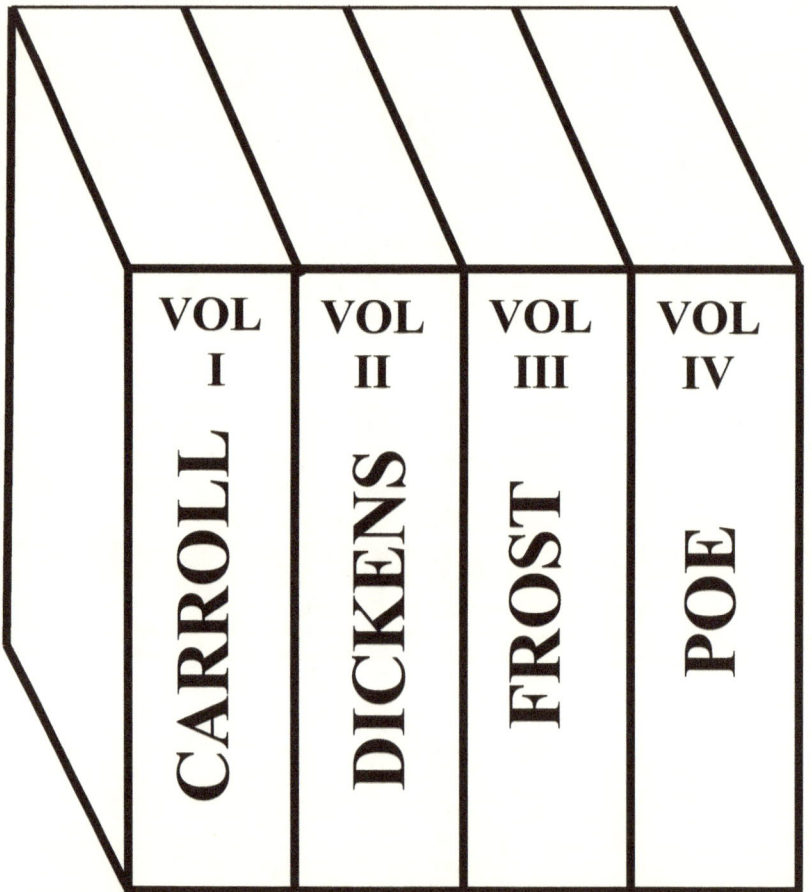

[Somewhere in a library, four volumes of literature are shelved as illustrated above.]

An erudite worm starts at the beginning of Volume I and eats his way to the end of Volume IV. If each of the covers is 1/8 of an inch thick, and the pages of each book have a total thickness of 2 inches, how far has this intelligent worm eaten?

Solution Option:

Instead of using arithmetic and logic, you may elect to solve the following elementary calculus integral:

The numerical value of this integral is equal to the number of inches eaten by the worm.

$$\int_{0}^{\sqrt[4]{19}} x^3 dx$$

!!

ANSWER: The bookworm has eaten through 4.75 inches of these four volumes.

2
A Logic Problem For The Ages

Question:

There is only one time in your life when you are twice as old as your child. When is that?

! !

The Solution will be presented both as a (1) process of logic, and a (2) mathematical process.

Solutions:

(1) When the child attains the same age that you (as the parent) were when the child was born, you will be twice as old as the child.

(2) Let A = age of the parent when the child is born.
   The age of the child when born is 0 years.
   When the child is A years old (A years later), the parent will be A + A = 2A years old.
   2A is twice A, so the parent will be twice as old as the child when the child is A years old.

3
Right Triangle Relationships

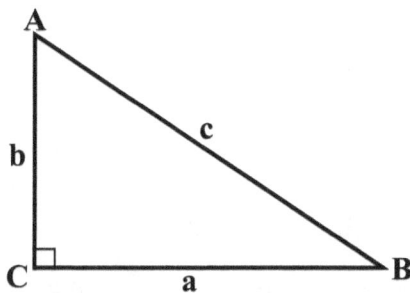

Consider the right triangle ABC with right angle C.
By the Pythagorean Theorem,

$c^2 = a^2 + b^2$, and
$c = \sqrt{a^2 + b^2}$

Question:
What is the relationship between (a + b) and c?
In other words, define this relationship mathematically.
Hint: There are exactly three possibilities, and only one is true.
! !
Solution:
(1) Assume that $(a+b) = c$
Then $(a+b) = \sqrt{a^2 + b^2}$
Therefore, $a^2 + 2ab + b^2 = a^2 + b^2$
This is <u>false</u> because $a^2 + 2ab + b^2$ is greater by $2ab$

(2) Assume that $(a+b) < c$
Then $(a+b) < \sqrt{a^2 + b^2}$
Therefore, $(a^2 + 2ab + b^2) < a^2 + b^2$
This is <u>false</u> because $a^2 + 2ab + b^2$ is greater by $2ab$

(3) Assume $(a+b) > c$
Then $(a+b) > \sqrt{a^2 + b^2}$
Therefore, $(a^2 + 2ab + b^2) > a^2 + b^2$ by $2ab$
This is TRUE.   (Q.E.D.)

4
Thinking Outside the Box

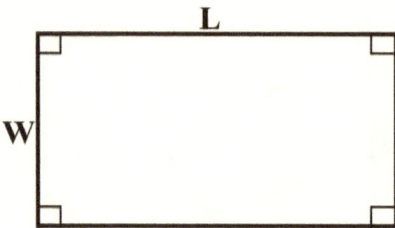

The area of a rectangle is usually expressed in terms of its length (L) and width (W). My challenge to you is to consider the unusual:

(I) Develop an equation to express the area of a rectangle, but do not use the width variable (W).

$$A = ?$$

(II) Develop an equation to express the area of a rectangle, but do not use the length variable (L).

$$A = ?$$

! !

Solution:

(1) $P = 2L + 2W$
   $(P = Perimeter)$
(2) $A = LW$

From (1): $L = \dfrac{P - 2W}{2}$, and $W = \dfrac{P - 2L}{2}$

Solution for (I):

$A = LW$

$A = (L)\left(\dfrac{P - 2L}{2}\right)$

$A = \dfrac{PL - 2L^2}{2}$

Solution for (II):

$A = LW$

$A = \left(\dfrac{P - 2W}{2}\right)(W)$

$A = \dfrac{PW - 2W^2}{2}$

Q.E.D.

5

Thinking Outside the Circle

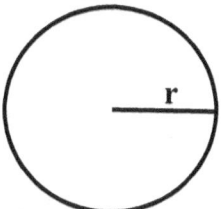

The circumference and area of a circle are usually expressed in terms of its radius or diameter. My challenge to you is to confront the extraordinary:

(I) Develop an equation that expresses the circumference of a circle in terms of the circle's <u>area</u>. In other words, the equation will <u>not</u> contain a variable for the radius or diameter.

$$C = ?$$

(II) Develop an equation that expresses the area of a circle in terms of the circle's <u>circumference</u>. In other words, the equation will <u>not</u> contain a variable for the radius or diameter.

$$A = ?$$

!!!

Solution:

(1) $C = 2\pi r$

(2) $A = \pi r^2$

From (1): $r = \dfrac{C}{2\pi}$

From (2): $r^2 = \dfrac{A}{\pi}$; $r = \sqrt{\dfrac{A}{\pi}}$; $r = \dfrac{\sqrt{\pi A}}{\pi}$

Therefore, (3): $\dfrac{C}{2\pi} = \dfrac{\sqrt{\pi A}}{\pi}$

Solution for (I):

In (3), solve for C:

$$C = 2\sqrt{\pi A}$$

Solution for (II):

In (3), solve for A:

$$C = 2\sqrt{\pi A}$$

$$C^2 = 4\pi A$$

$$A = \frac{C^2}{4\pi}$$

Q.E.D.

6
Head Travel

Suppose you went on a long walking tour around the earth's equator, and suppose that you are six feet tall. (For the purpose of friendlier numbers for calculations, assume that the earth is a perfect sphere and the distance around the equator is 25,000 miles.)

Question:
How much farther would your head travel than your feet?
!!!

Answer:
Perhaps you are surprised that your head travels farther than your feet!

Reference the diagram:
As your feet travel along the inner circle (in contact with the earth's surface), your head will obviously travel along the outer circle.

$C = 2\pi R$ represents the circumference of the earth at the
      equator = 25,000 miles
C_F = feet circle (path of feet)
C_H = head circle (path of head)

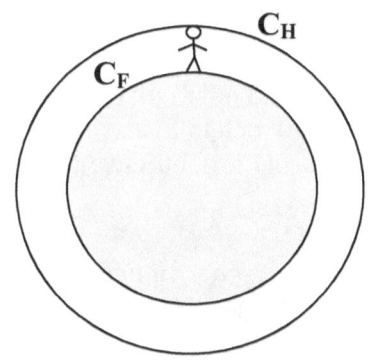

$C_F = 2\pi R$

$C_H = 2\pi\left(R + \dfrac{6}{5280}\right)$

$C_H = 2\pi R + \dfrac{12\pi}{5280}$

$C_H = 25000 + \dfrac{12\pi}{5280}$

$C_H = 25,000.00714$ miles

$(.00714)(5280) = 37.7$ feet

To the nearest tenth of a foot, your head will travel 37.7 feet farther in 25,000 miles.
This is nothing to worry your head about, but regardless of how far you walk,
your head will always travel farther than your feet. ☺

7

Traveling in Circles

Assume that two cars are both traveling with speeds of 70 mph on roads around the equator of a perfectly spherical earth whose diameter is 8000 miles. The red car is traveling on a road directly on the equator; the blue car is traveling on a road elevated 15 feet above the equator for the entire distance around the earth.

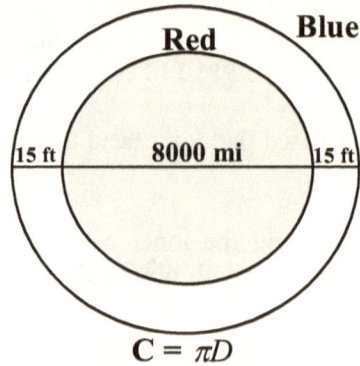

Question:
How much longer will be required for the blue car to finish one revolution?
! !
Answer:
Convert 8000 mi. to ft.: (8000 mi.)(5280 ft./mi.) = 42,240,000 ft.
Convert 70 mph to ft./sec.:
(70 mi./hr.)(1 hr./3600 sec.)(5280 ft./mi.) = 102.67 ft./sec.

$C_{Red} = \pi D$
$\qquad = (3.14159265)(42,240,000 \text{ ft.})$
$\qquad = 132,700,874 \text{ ft.}$

$D = RT$
$T = D/R$
$T = 132,700,874 \text{ ft.}/102.67 \text{ ft./sec.}$
$T = 1,292,499.016 \text{ sec.}$

$C_{Blue} = \pi D$

$\qquad = (3.14159265)(42,240,030 \text{ ft.})$

$\qquad = 132,700,968 \text{ ft.}$

$D = RT$

$T = D/R$

$T = 132,700,968 \text{ft.}/102.67 \text{ ft./sec.}$

$T = 1,292,499.932 \text{ sec.}$

Time Difference $= 1,292,499.932 \text{ sec.} - 1,292,499.016 \text{ sec.}$

$\qquad\qquad\qquad = .916 \text{ sec.}$

The Blue car finishes less than 1 second behind the Red car.

8
Cubical Water Tank

7.48052 gal/ft^3

Drain Rate:
1 gal/sec

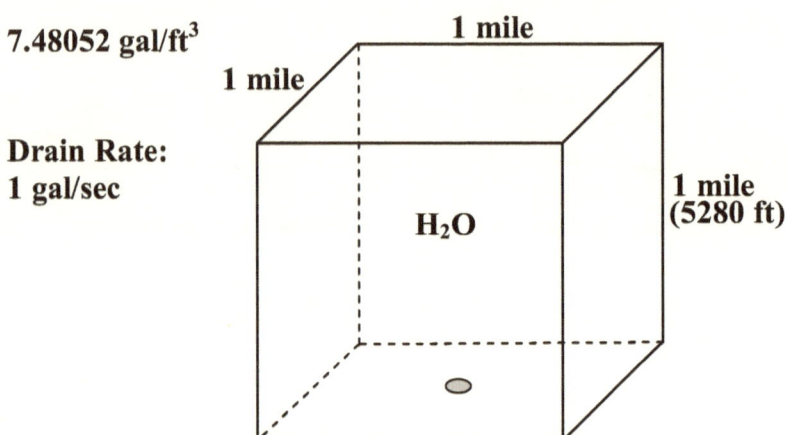

A cubical water tank has an edge equal to 1 mile (5280 ft.), and is completely filled with water. In the middle of the bottom side of the tank is a drain plug. If opened, the water will drain at the rate of 1 gallon per second.

Question:

What time will be required to completely empty the tank? Express your answer to the nearest whole year.

! !

Solution:

$(5280 \text{ ft.})^3 = 1.47197952 \times 10^{11} \text{ft.}^3$

There are 7.48052 gal./ft.^{3}

$(1.47197952 \times 10^{11} \text{ft.}^3)(7.48052 \text{ gal./ft.}^3)$

$= 1.101117224 \times 10^{12}$ gal.

Draining at the rate of 1 gal./sec:

$1.101117224 \times 10^{12}$ gal. $= 1.101117224 \times 10^{12}$ sec.

For 365 days/yr.:

$$\frac{1.101117224 \times 10^{12} \text{ sec.}}{31,536,000 \text{ sec./ } yr.} = 34,916.19182 \, yr. \approx 34,916 \, yr.$$

For 365.25 days/yr.:

$$\frac{1.101117224 \times 10^{12} \text{ sec.}}{31,557,600 \text{ sec./ } yr.} = 34,892.29929 \, yr. \approx 34,892 \, yr.$$

108

9
A COIN PROBABILITY PROBLEM

A box contains two coins. One coin has "heads" on both sides. The other coin has "heads" on one side and "tails" on the other side. All three "heads" have <u>identical</u> configurations. A coin is selected randomly. The observed side of the selected coin is "heads".

Question:
What is the probability that the other side of the selected coin is <u>also</u> "heads"?
!!
Answer:
The probability is Two Chances out of Three (2/3 or $66\frac{2}{3}$ %).

Suppose you label the double-headed coin "heads #1" on one side and, "heads #2" on the other side. The regular coin is simply "heads" and "tails".
You have an equal chance of observing any of the following sides:

      (a) heads
      (b) tails
      (c) heads #1
      (d) heads #2

In this problem, you observe one of the heads. This eliminates the (b) tails possibility. (You did not observe tails).
Three possibilities remain:

      (a) heads (with tails on the reverse side)
      (c) heads #1 (with heads #2 on the reverse side)
      (d) heads #2 (with heads #1 on the reverse side)

Therefore, the chances are two out of three that the other side of the coin is one of the heads.

10
A LOGIC/PROBABILITY PROBLEM

Four identical sealed envelopes are on a table. One of them contains a $100 bill. You randomly select an envelope and hold it in your hand without opening it.

Two of the three remaining envelopes are then removed and set aside, unopened. You are told that these two envelopes are empty (they truthfully are empty).

You are given the choice of keeping the envelope you chose or exchanging it for the one on the table.

What should you do?
        (A) Keep your envelope
        (B) Switch it
        (C) It does not matter
!!
Answer:
The correct answer is (B).
This choice will give you the best probability of acquiring the $100.
Following is the reasoning for this conclusion:
When you randomly choose your envelope, your probability of success is 25% (1 chance out of 4 that your envelope contains the $100).

This also logically means that the probability of success for the remaining group of three envelopes must be 75% (3 chances out of 4 that one of these envelopes contains the $100).

The fact that 2 of the 3 remaining envelopes are removed does not alter the success probability of 75% for the group of 3 envelopes. The fact that the 2 removed envelopes are <u>empty</u> only increases the confidence that the probability that the remaining envelope more likely contains the $100 than the envelope you randomly selected.

11
PROOF THAT 1 = 2

1. *Let* $A = B$

2. Square both sides of the equation (Multiply both sides by *B)*:
   $$(A)(B) = B^2$$

3. Subtract A^2 from both sides of the equation:
   $$(A)(B) - A^2 = B^2 - A^2$$

4. Factor both sides of the equation:
   $$A(B - A) = (B + A)(B - A)$$

5. Divide both sides of the equation by $(B - A)$:
   $$A = (B + A)$$

6. By the authority of step #1, if $A = 1$, *then* $B = 1$:
   $$1 = 1 + 1$$

7. Therefore, $1 = 2$

Q.E.D.

Postscript:
Can you determine the error in this proof, or does your indifferent sensibility result in your contentment to accept its invalidity?

! !
ANSWER:
The error is in step 5.
$(B - A)$ **is 0. You cannot divide by 0. If you do, bad and unpredictable results occur, like proving that 1 = 2.** ☹

12
Inscribed & Circumscribed

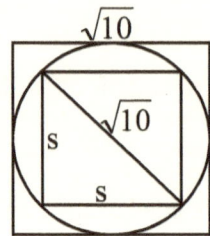

In the illustration above:

From an inside to outside perspective:
The small square is inscribed in the circle.
The circle is inscribed in the large square.
The larger square has an area of 10 square units.

From an outside to inside perspective:
The larger square is circumscribed about the circle.
The circle is circumscribed about the smaller square.
The larger square has an area of 10 square units.

Question:
What is the area of the smaller square?
!!!
Answer:
Because the area of the larger square is 10 square units, each side is $\sqrt{10}$.
The smaller square diagonal is equal to the circle diameter.
The circle diameter is equal to the side of the larger square.

Therefore, the diagonal of the smaller square is $\sqrt{10}$.

Let s = the side of the smaller square.
Then:

$$s^2 + s^2 = \left(\sqrt{10}\right)^2$$
$$2s^2 = 10$$
$$s^2 = 5$$

The area of the smaller square is 5 square units.
Q.E.D.

13
Tissue Paper Thinking

Imagine a steel band fitting tightly around the equator of the earth (assume that the earth is a perfect sphere). Now, suppose that you remove it and then splice in an additional piece 10 feet long, so that the new band is 10 feet longer than the original band. If you now replace it on the equator, it would fit more loosely all around the circumference of the earth.

Question:
How great a distance would there now be between the band and the earth?
Would the distance be sufficient for:
(a) A person, 6 feet tall, to walk through?,
(b) A person to crawl through on hands and knees?,
(c) A piece of tissue paper to just slip through?, or
(d) None of the above?
!!

Answer:
Reference the diagram illustrating the two different circumferences:

C_E = Circumference of earth
C_B = Circumference of new band having 10 additional feet
$C_E = 2\pi R$
$C_B = 2\pi(R+x)$, where x is the increase in the radius
$C_B = 2\pi R + 2\pi x$
$C_B = C_E + 10$ feet
$C_E + 10$ feet $= 2\pi R + 2\pi x$

A 10 feet increase in the circumference results in a $2\pi x$ increase in the radius.
Therefore:
$10 = 2\pi x$
$$x = \frac{10}{2\pi}$$
$x \approx 1.59$ feet

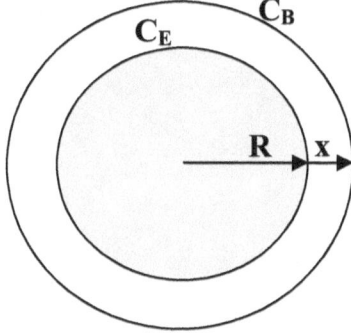

So, the radius of the new band is approximately 1.6 feet. Yes, the new band will be about 1.6 feet above the equator, <u>all around the equator</u>.

Therefore, the answer to the question is (b). ☺

Q.E.D.

14
An Algebra Problem With A Temperature

At high noon on October 16, 2012 in Clearwater, Florida and Lafayette, Indiana, a total of 121 degrees was the sum of their 2-digit Fahrenheit temperatures.

The tens' digit of the Clearwater temperature was three greater than its units' digit. If you reversed the two digits of the Clearwater temperature, the result was the Lafayette temperature.

Question:
What were the temperatures of these two cities?

Anyone can guess at anything; however, for this problem:
You do the math!
        and
You show the math!
!!
Solution:
Let t = the tens' digit of the Clearwater temperature.
Then $t - 3$ = the units' digit.
$10(t) + 1(t - 3)$ = Clearwater temperature.
Reversing the digits for the Lafayette temperature:
$10(t - 3) + 1(t)$ = Lafayette temperature.
The sum of the two temperatures is 121 degrees:
$10t + t - 3 + 10t - 30 + t = 121$.
$22t - 33 = 121$.
$22t = 154$.
$t = 7$.
$t - 3 = 4$.
The Clearwater temperature was 74 degrees.
The Lafayette temperature was 47 degrees.
Q.E.D.

15
A CARD PROBABILITY PROBLEM

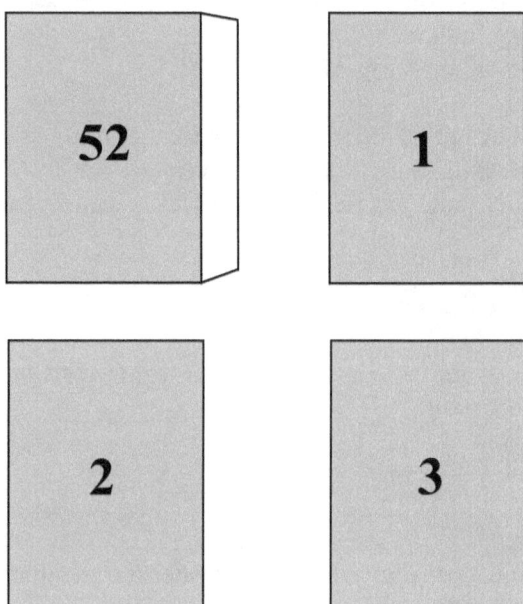

An ordinary deck of 52 playing cards (already thoroughly shuffled) is shuffled completely (this means the deck is shuffled and repeatedly reshuffled, cut, re-cut, reshuffled, cut, re-cut, etc.). You may assume that the card deck is in a "theoretical random" sequence. That is, the sequence is not biased in any manner. Three cards are sequentially removed from the top of the deck and placed face down. Card 1 is removed first; Card 2 is the second card removed; Card 3 is the third card removed.

There are four probability questions. Express each of the four probabilities as a Decimal Fraction <u>rounded</u> to four decimal places (e.g., .XXXX), and as a Per Cent <u>rounded</u> to two decimal places (e.g., XX.XX%).

!!

Solution:

(1) What is the probability that NONE of the three cards is a Red Queen?

(50/52) x (49/51) x (48/50) = .8869 = 88.69%

Suppose that <u>only</u> <u>one</u> of these three cards IS a Red Queen. <u>Which</u> <u>card</u> is <u>most</u> <u>probably</u> the Red Queen? The answer can be determined after answering the next three questions.

(2) What is the probability that the FIRST card is a Red Queen?

2/52 = .0385 = 3.85%

(3) What is the probability that the FIRST card is NOT a Red Queen, AND the SECOND card IS a Red Queen?

(50/52) x (2/51) = .0377 = 3.77%

(4) What is the probability that the FIRST card is NOT a Red Queen, AND the SECOND card is NOT a Red Queen, AND the third card IS a Red Queen?

(50/52) x (49/51) x (2/50) = .0370 = 3.70%

16
Poetic Equation

Question:

Can you read the following equation as a poem?

$$\frac{12 + 144 + 20 + 3\sqrt{4}}{7} + (5 \times 11) = 9^2 + 0$$

! !

Answer:

The answer is a poem having four lines with a rhyming pattern of aaba:

A dozen, and a gross, and a score,
Plus three times the square root of four;
Divide it by seven, plus five times eleven
Equals the square of nine plus nothing more.

Q.E.D.

17
A MARBLES LOGIC PROBLEM

An urn contains exactly 100 spherical glass marbles having the same smooth texture. There are five different colors. The color distribution is as follows:

    30 Blue marbles

    30 Red marbles

    30 Green marbles

    The 10 remaining marbles are an unknown mixture of White and Yellow.

The urn is shaken thoroughly and a marble is selected unseen. This procedure is repeated for each marble selected.

Question:
What is the <u>Absolute</u> <u>Minimum</u> number of marbles that must be selected in order to <u>Assure</u> with <u>Absolute</u> <u>Certainty</u> that 10 marbles of the <u>same</u> color have been selected?
!!
Answer:
Consider the worst case scenario: In whatever order, you manage to remove 9 of the blue marbles, 9 of the red marbles, 9 of the green marbles, and all 10 of the white and yellow mixture. So far, you have removed 37 of the marbles (9 + 9 + 9 + 10 = 37), but you do not have 10 of any single color. If you remove one more marble, it will be blue or red or green, and you will have 10 of the same color. Therefore, the correct answer is 38 marbles.

In other words, 38 is the lowest possible number, given the worst case scenario, to be sure that at least 10 marbles are the same color.

In other words again, it is necessary to select exactly 38% of the original marbles in order to be assured that 10 marbles have the same color.

18
DIE PROBABILITY PROBLEM

Question:
Suppose you plan an experiment in which you will roll a die 20 times. Which of the following results is more likely?

(A) 11111111111111111111

OR

(B) 66234441536125563152
!!
Answer:
Because the experiment has not yet occurred, the correct answer is that both results (A) and (B) are <u>equally</u> likely.
However, suppose that, out of your view, I rolled a die 20 times and noted the result, digit by digit. Then, I informed you that my result was either:
(A) 11111111111111111111

OR

(B) 43335643531612115412

Which of these two results is more likely? The correct answer is now (B) because the 20 rolls have already occurred, and the result is far more likely to have been the mixture of numbers in (B) rather than a series of twenty ones. Why? It is because the probability of getting result (A) is only one chance out of

3,656,158,440,062,976 results, or $\dfrac{1}{3,656,158,440,062,976}$.

The chance of getting <u>any</u> <u>other</u> sequence of numbers is 3,656,158,440,062,975 chances out of 3,656,158,440,062,976, or $\dfrac{3,656,158,440,062,975}{3,656,158,440,062,976}$.

So, the probability of (B) occurring is <u>very</u> close to 1*. This is why, after the experiment has <u>already</u> occurred, that it is extremely unlikely that (A) has occurred.

*(A probability of 1 means that a particular result <u>will</u> occur).

Q.E.D.

Reference the next page Appendix for the number of possible outcomes of rolling a die from one to twenty times.

DIE PROBABILITY PROBLEM

APPENDIX

6^n

Determine 6^1 thru 6^{20}

n

1 6
2 36
3 216
4 1,296
5 7,776
6 46,656
7 279,936
8 1,679,616 [10^6]
9 10,077,696
10 60,466,176
11 362,797,056
12 2,176,782,336 [10^9]
13 13,060,694,016
14 78,364,164,096
15 470,184,984,576
16 2,821,109,907,456 [10^{12}]
17 16,926,659,444,736
18 101,559,956,668,416
19 609,359,740,010,496
20 3,656,158,440,062,976 [10^{15}]

19
Basket of Eggs

A country grocery store clerk prepares a basket of fresh eggs ordered and purchased by a customer. The customer comes to the store and tells the clerk:

"I'd like half the eggs in the basket plus half an egg, please."

The clerk gives the requested number of eggs to the customer. The following day, the customer returns and instructs the clerk:

"I'd like half the eggs in the basket plus half an egg, please."

Again the clerk provides the requested number of eggs for the customer. The next day. the customer returns again and tells the clerk:

"I'd like half the eggs in the basket plus half an egg, please."

The clerk gives the requested number of eggs to the customer, and the basket is now <u>empty</u>.

Question:

How many eggs were originally in the basket?

!!

Solution:

Let N = the number of eggs in the basket originally.

Determine the number of eggs taken the first day:

$$\frac{N}{2}+\frac{1}{2}=\frac{N+1}{2}$$

Determine the number of eggs remaining after the first day:

$$N-\left(\frac{N+1}{2}\right)=\frac{N-1}{2}$$

Determine the number of eggs taken the second day:

$$\left(\frac{\frac{N-1}{2}}{2}\right)+\frac{1}{2}=\left(\frac{N-1}{4}\right)+\frac{1}{2}=\frac{N+1}{4}$$

Determine the number of eggs remaining after the second day:

$$\left(\frac{N-1}{2}\right)-\left(\frac{N+1}{4}\right)=2\frac{(N-1)-(N+1)}{4}=\frac{N-3}{4}$$

Determine the number of eggs taken the third day:

$$\left(\dfrac{\dfrac{N-3}{4}}{2}\right)+\dfrac{1}{2}=\left(\dfrac{N-3}{8}\right)+\dfrac{1}{2}=\dfrac{N+1}{8}$$

The sum of all the eggs taken for the three days equals the number of eggs that were originally in the basket:

$$\left(\dfrac{N+1}{2}\right)+\left(\dfrac{N+1}{4}\right)+\left(\dfrac{N+1}{8}\right)=N$$

$$4N+4+2N+2+N+1=8N$$

$$7N+7=8N$$

$$N=7\ eggs$$

Answer Check:

Eggs taken the 1st day:

$$\dfrac{7}{2}+\dfrac{1}{2}=4\ eggs$$

Eggs remaining after 1st day:

$$7-4=3\ eggs$$

Eggs taken the 2nd day:

$$\dfrac{3}{2}+\dfrac{1}{2}=2\ eggs$$

Eggs remaining after the 2nd day:

$$3-2=1\ egg$$

Eggs taken the 3rd day:

$$\dfrac{1}{2}+\dfrac{1}{2}=1\ egg$$

Eggs remaining after the 3rd day:

$$1-1=0\ eggs$$

Q.E.D.

20
Sphere in a Cube

Given:

A sphere is inscribed in a cube (which also means that the cube is circumscribed about the sphere).

The diagonal (AB) of the cube face = $3\sqrt{6}$ *meters*.

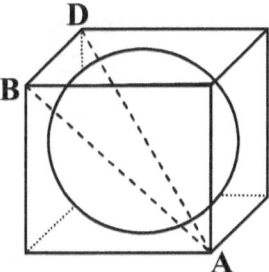

Use a calculator for the calculations involved in the following questions. There are ten questions. Each question is worth ten points. The total number of points is 100. 100% represents Success. ☺; any other percentage is Failure. ☹

What is the volume of the sphere?

> ➤ Use the symbol "π" as π in the answer. Do not convert π to any numerical value. Simplify the expression of the answer as much as possible. If any part of the answer cannot be converted to a rational number, leave it as a radical (do <u>not</u> convert the radical to a numerical value).

(1) $\left(\dfrac{27\sqrt{3}}{2}\right)\pi$ (2) *meters*³

  (calculated value   (unit of measurement)
                      (same for (1) and (3))

> ➤ Use the calculator value of π, completely calculate the expression of (1) including any radical, and round the completely numerical answer to <u>two</u> decimal places.

(3)   73.46
     (calculated value)

What is the ratio of the spherical volume to the cubical volume?

> ➤ Use the calculator value of π, round the answer to <u>two</u> decimal places, and multiply by 100 to express the answer as a percent.

(4) 52%
(calculated percent)

What is the ratio of the surface area of the sphere to the surface area of the cube?

> ➤ Use the calculator value of π, round the answer to two decimal places, and multiply by 100 to express the answer as a percent.

(5) 52%
(calculated percent)

What is the circumference of the sphere?

> ➤ A plane passed thru the center of the sphere describes a great circle on the sphere's surface. This circle represents the perimeter or circumference of the sphere.

> ➤ Use the calculator value of π, and round the answer to <u>two</u> decimal places.

(6) 16.32 *meters*
(calculated value)

What is the area of the planar area ABD?

> ➤ This triangular area is formed by lines AB, BD, and AD.

> ➤ Round the answer to <u>two</u> decimal places.

(7) 19.09   **(8)** *meters*^{2}
(calculated value)   (unit of measurement)

What is the length of the cube diagonal AD?

➤ Round the answer to <u>two</u> decimal places.

(9) 9.00 *meters*
(calculated value)

How many points of tangency (contact points) are there between the sphere and the cube?

➤ Express the answer as an integer.

(10) 6

21
Precision Design

As illustrated in Diagram 1, a perfect sphere is inscribed in a perfect cube. The dimensions of both are precisely designed so that the sphere and cube are tangent to each other. This indicates that the sphere and cube contact each other at only six points – the exact centers of each cube face. An edge of the cube has a length of "D". A second, smaller sphere, is positioned in the front right corner of the cube. This sphere is tangent to the large sphere, and tangent to the cube on the right, front, and bottom faces. This indicates that the small sphere contacts the large sphere at only one point, and the small sphere contacts three cube faces – right, front, and bottom, for a total of four contact points. The cube's diagonal from the top left rear vertex to the bottom right front vertex, intersects the centers of each sphere, and passes through the tangential contact of the two spheres.

As illustrated by Diagram 2, the small sphere is inscribed in a small cube such that both are tangent to each other. An edge of the small cube has a length of "d". The small sphere is the same sphere that is illustrated in Diagram 1. Obviously, the small cube of Diagram 2 cannot possibly reside in the large cube. It is illustrated here to define the dimension of the small sphere, and to illustrate the fact that the small cube diagonal is indeed a segment of the large cube diagonal.

The Challenge

(1) Determine the equation that will express the small sphere diameter in terms of the large sphere diameter.

(2) After determining the appropriate equation, find the value of the small sphere diameter to the nearest millionth of a unit when the large sphere diameter is equal to 1 unit.

Solution for (1):

$$\frac{D}{2} + \frac{d}{2} + \frac{d\sqrt{3}}{2} = \frac{D\sqrt{3}}{2}$$

$$D + d + d\sqrt{3} = D\sqrt{3}$$

$$d = \frac{D(\sqrt{3}-1)}{\sqrt{3}+1}$$

Solution for (2):

Let $D = 1$ unit

$$d = \frac{D(\sqrt{3}-1)}{\sqrt{3}+1}$$

$$d = \frac{\sqrt{3}-1}{\sqrt{3}+1}$$

$$d = .267949 \text{ unit}$$

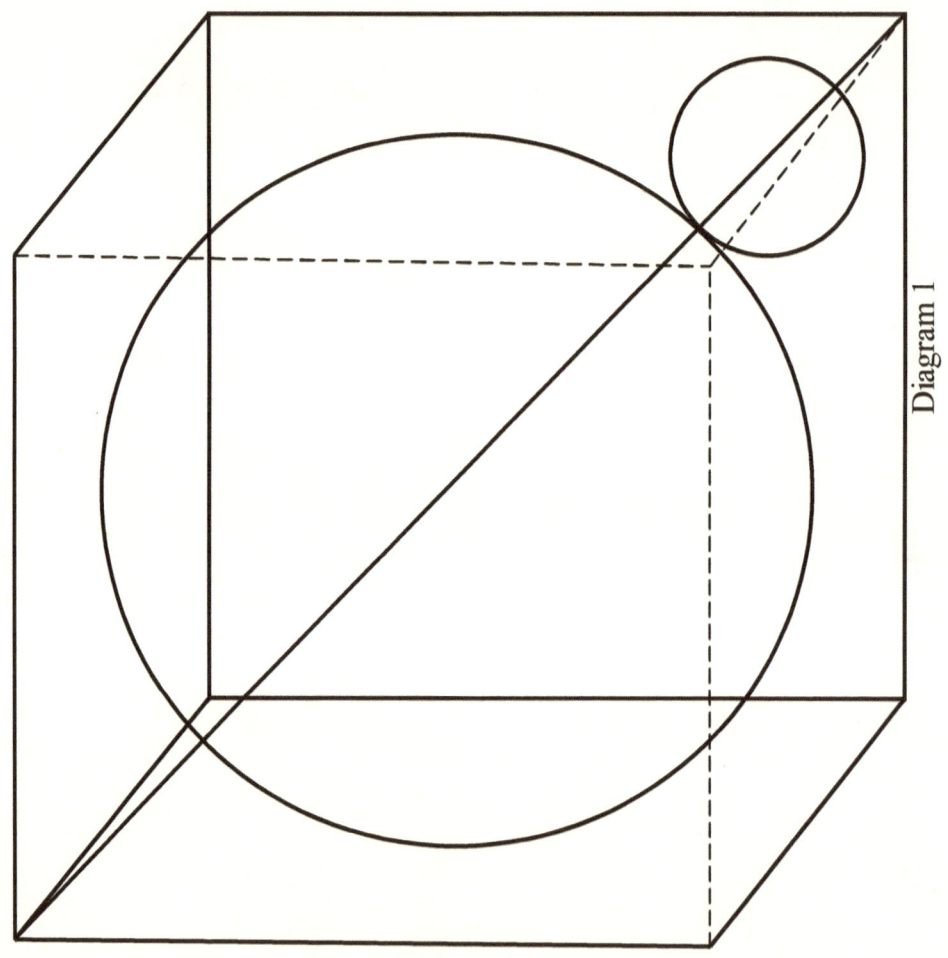

Diagram 1

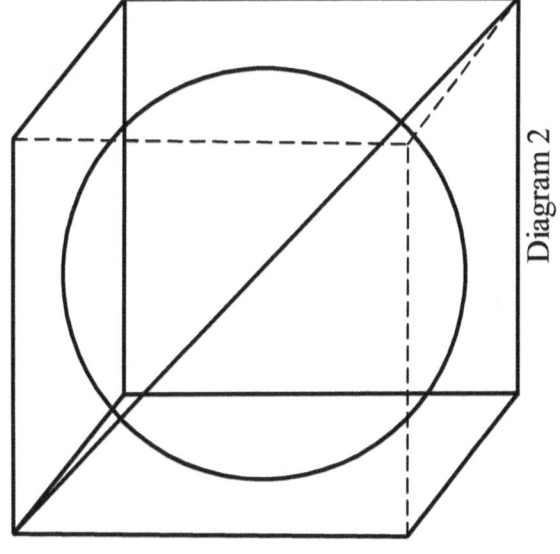

Diagram 2

22

Algebra – Fundamental Operations
EXAM

Simplify (a), (b), (c), and (d).
Then perform the operations specified by (e), (f), and (g).
All variables are alphabetic characters.
After operation (g), <u>rearrange the eleven alphabetic characters</u> to form a three-word phrase. ☺

(a)
$$\frac{J\left[\left(E^{\frac{1}{2}}R^{\frac{1}{2}}Z-A\right)\left(E^{\frac{1}{2}}R^{\frac{1}{2}}Z+A\right)(S)^{2}+S^{2}A^{2}\right]}{RSZ^{2}}$$

(a) = JES

(b)
$$\frac{\dfrac{U+S}{US}+\dfrac{S}{U^{2}-US}}{\dfrac{U}{S^{2}(U^{2}-US)}}$$

(b) = US

(c)

$$EY\left(\frac{4IS}{3\left((E+Y)^{2}-(E-Y)^{2}\right)}\right)+ZY\left(\frac{4IS}{3\left((Y+Z)^{2}-(Y-Z)^{2}\right)}\right)$$

$$+EZ\left(\frac{4IS}{3\left((Z+E)^{2}-(Z-E)^{2}\right)}\right)$$

(c) = IS

(d)
$$\frac{2L\sqrt{C^{2}Z^{2}\sqrt[3]{O^{6}}}}{3BZ}\times\frac{3\sqrt[3]{B^{3}Z^{3}\sqrt{(RD)^{6}}}}{2CZ}$$

(d) = LORD

132

(e) Multiply (a) and (b).   **(e) = JESUS**

(f) Add (c) to (e).   **(f) = JESUS + IS**

(g) Add (d) to (f).   **(g) = JESUS + IS + LORD**

Yes, JESUS IS LORD!!!

23
Mathematics – Fundamental Algebra
EXAM

(1)

Perform the indicated operations and simplify:

$$\frac{\dfrac{a-b}{a}-\dfrac{b}{a+b}}{\dfrac{a+b}{b}-\dfrac{a}{a-b}}+\dfrac{1-\dfrac{b}{a}}{\dfrac{a}{b}+1} \qquad \textbf{(1)} \quad \frac{2b(a-b)}{a(a+b)}$$

(2)

Solve the system of simultaneous linear equations for x, y, and z:

$$\frac{3x}{2}-y+\frac{2z}{3}=4$$

$$\frac{x}{3}+\frac{5y}{2}-\frac{z}{6}=\frac{8}{3} \qquad \textbf{(2)} \quad (x=2,\ y=1,\ z=3)$$

$$\frac{x}{6}-\frac{2y}{3}+\frac{z}{9}=0$$

(3)

Perform the indicated operations and simplify:

$$\frac{y^{-2}\left(x^{3}y^{-1}+x^{2}\right)}{x^{2}\left(1+x^{-1}y\right)} \qquad \textbf{(3)} \quad \frac{x}{y^{3}}$$

(4)

Solve the inequality for x, then graph the solution set:

$$\left(\frac{x+8}{4}-1\right)>\frac{x}{3} \qquad \textbf{(4)} \quad x<12$$

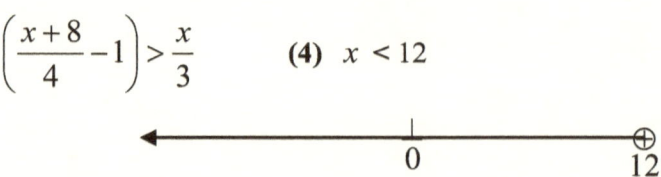

(5)
In aerospace engineering, the radius (r) of a satellite orbit is determined by the formula below, where t represents the time required for the satellite to complete one orbit, G represents the universal gravitational constant, and M represents the mass of the central body (earth). Solve the formula for t:

$$r = \sqrt[3]{\frac{GMt^2}{4\pi^2}}$$

$$(5) \quad t = \left(\frac{2\pi r}{GM}\right)\sqrt{GMr}$$

(6)
The following proof is presented for your analysis. The conclusion, obviously wrong, is logically consistent from the sequence of mathematical operations; however, one of the steps is <u>invalid</u>. Your objective is to determine which step (other than step 7) is invalid, and why.

PROOF THAT $1 = 2$
1. Let $A = B$
2. Square both sides of the equation:
$$AB = B^2$$
3. Subtract A^2 from both sides of the equation:
$$AB - A^2 = B^2 - A^2$$
4. Factor both sides of the equation:
$$A(B - A) = (B + A)(B - A)$$
5. Divide both sides of the equation by $(B - A)$:
$$A = B + A$$
6. From Step 1, if $A = 1$, then $B = 1$:
$$1 = 1 + 1$$
7. Therefore, $1 = 2$

Question 1: Which step is invalid? **5.**
Question 2: Why is the step invalid? **Division by 0 is undefined.**

(7)
According to current Florida State Law, the sales tax on a new automobile is calculated as follows:
The first $5000 of the sale price is taxed at the rate of 7%.
The balance of the sale price is taxed at the rate of 6%.
Let S represent the Sale price of an automobile. Write the equation, in terms of S, to calculate the Total Sale price (including the tax). Express the equation in simplest form. Do not substitute some specific numerical value for S. The objective is to determine an algebraic equation that is a general solution.

(7) TS = S + .06S + 50 or TS = 1.06S + 50

(8)
Assuming that fractional hens are as proportionally functional as whole hens, consider the following question:
If a hen-and-a-half lay an egg-and-a-half in a day-and-a-half, how many eggs will three hens lay in three-and-a-half days?
To avoid possible confusion, the same question will be expressed differently:
If 1.5 hens lay 1.5 eggs in 1.5 days, how many eggs will 3 hens lay in 3.5 days?

Hens	Eggs	Days
3/2	3/2	3/2
3	**3**	**3/2**
3	7	7/2

(9)
Suppose the earth is a perfect sphere, and there is a steel band fitting tightly around the equator. Now suppose that you remove the band and cut it at one place, then add an additional piece that is 10 feet long. The new band is now 10 feet longer than the original band. If you replace the band on the equator, it will fit more loosely. It will fit more loosely because an increased circumference also increases the radius. Previously, there was no space between the band and the earth, but now the increased radius causes a space between the band and the earth all around the equator. (Reference the diagram).

Let C = circumference of the original band in feet.
Let R = radius of the original band in feet.
Let r = the increase of the radius in feet (only the increase, not the entire radius of the new band).
Let π = 3.14

There are two questions:
(1) What is r, to the nearest <u>tenth</u> of a foot?   **1.6 feet**
(2) Select the answer below that is most logically consistent
     with the value of r.

>   The space (r) between the new band and the earth will be
>   great enough for:
>>   ___(a) a person, 6 feet tall, to walk through, or
>>   ✓ (b) a person to crawl through, or
>>   ___(c) a piece of tissue paper to tightly slip
>>   through, or
>>   ___(d) none of the above?

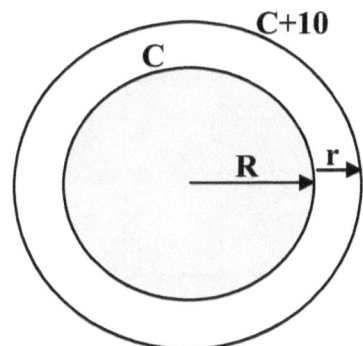

(10)
Suppose that a singles tennis tournament has been planned. It
will be a double-elimination tournament (a participant is
eliminated after losing twice). The number of participants will be
N. One can of tennis balls will be provided for each and every
tennis match required. Expressed in terms of N, how many cans
of tennis balls are necessary for this tournament? (Because the
number of participants is N, do not express the answer with a
specific numerical value for the number of cans. The answer is
an algebraic expression for a general solution).

(10) 2(N – 1) or 2(N – 1) + 1

24
Up Against A Brick Wall 1

A large, square, brick wall (wall #1) is one brick higher and one brick narrower than another wall (wall #2). **How many <u>more</u> bricks are in wall #1?** [Assume that all the bricks are square]. You can obviously count the bricks in the illustration.
Develop an algebraic formula to prove the general solution.

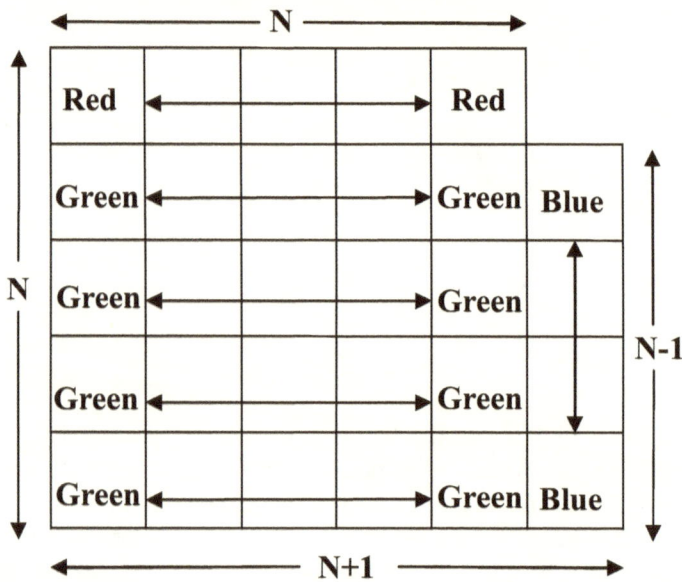

The square bricks that are Red and Green represent the square Wall #1.
The square bricks that are Green and Blue represent Wall #2.
The Green bricks are common to both walls.
!!!
Solution:
Let N represent the number of bricks in the height and width of Wall #1.
Then N-1 represents the number of bricks in the height of Wall #2, and N+1 represents the number of bricks in the width of Wall #2.

Therefore, the number of bricks in Wall #1 = N x N = N^2.
Further, the number of bricks in Wall #2 = (N-1)(N+1)
$$= N^2 - 1.$$

138

So, Wall #1 obviously has <u>one more</u> brick than Wall # 2 because N^2 is 1 more than $N^2 - 1$.☺

Note: If the bricks are not square, and therefore neither wall is square, then what is the answer?
"Up Against A Brick Wall 2" presents the problem with rectangular bricks (the height and width are not equal).

25
Up Against A Brick Wall 2

Consider an alternative problem regarding the two brick walls having the condition that all bricks are rectangular (not square). This is actually the normal situation. The dimensions of height (h) and width (w) of each brick are unequal, but all the bricks are exactly the same size.

Now, how many bricks does one wall have relative to the other wall?

Develop an algebraic formula to prove the general solution.

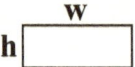

w > h for each brick; The front surface area of each brick = hw

Wall #1 Bricks are Red and Green (the same color pattern as "Up Against A Brick Wall 1").

Wall #2 Bricks are Blue and Green (the same color pattern as "Up Against A Brick Wall 1").

The Green bricks are common to both walls.

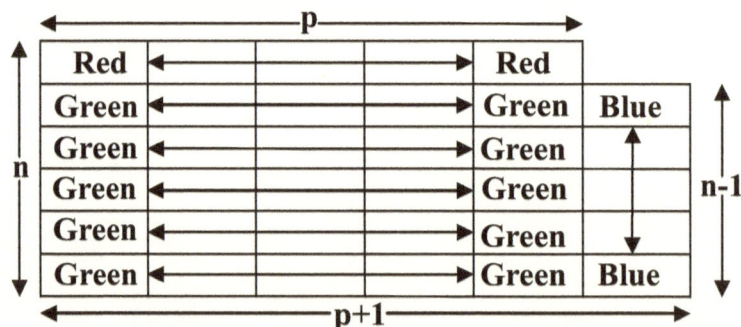

!!!

Solution:

Area of Wall #1:

Let n represent the number of bricks in the height of Wall #1.

Let p represent the number of bricks in the width of Wall #1.

The height of Wall #1 is nh, and the width of Wall #1 is pw.

Therefore, the area of Wall #1 is (pn)(hw).

Up Against A Brick Wall 2

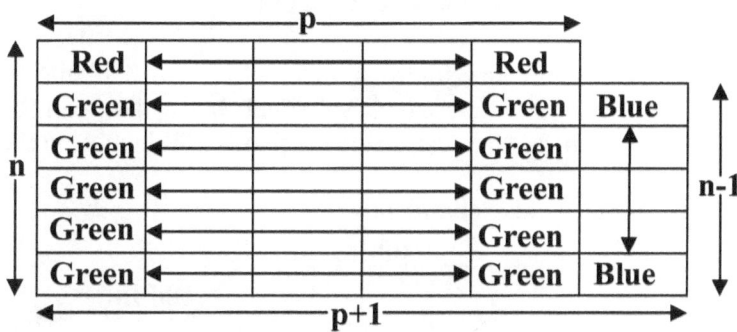

Area of Wall #2:

The height of Wall #2 is (n-1)h, and the width of Wall #2 is (p+1)w.

The area of Wall #2 (the product of the height and width) is [(n-1)h][(p+1)w].

[(n-1)h][(p+1)w] = (nh-h)(pw+w).

(nh-h)(pw+w) = (pn)(hw)-nhw-phw-hw.

Therefore, the area of Wall #2 is (nh-h)(pw+w)

= (pn)(hw)+(n-p-1)hw.

From the last equation above, n-p-1 = N

(Number of bricks in Wall #2 <u>relative</u> to Wall #1).

If n-p = 1, both walls have the same number of bricks.

If n-p = 0, Wall #2 has one <u>less</u> brick.

If n-p = 2, Wall #2 has one <u>more</u> brick.

If n-p = -1, Wall #2 has two <u>less</u> bricks.

If n-p = 3, Wall #2 has two <u>more</u> bricks.

If n-p = -2, Wall #2 has three <u>less</u> bricks.

If n-p = 4, Wall #2 has three <u>more</u> bricks.

Substitute any values of n and p to determine the # of bricks in Wall #2 relative to Wall #1.

Note: The relative number of bricks between Wall #1 and Wall #2 does <u>not</u> depend on the h and w values of the bricks; the dependence is only on n (the number of bricks in the vertical height of Wall #1) and p (the number of bricks in the horizontal width of Wall #1).

26
Box of Balls

Question:

What is the absolute maximum number of one-inch diameter, non-compressible stainless steel spherical balls that can be packed into a cubical stainless steel box that is 100 inches on the inside edge?

(The box top must close completely).

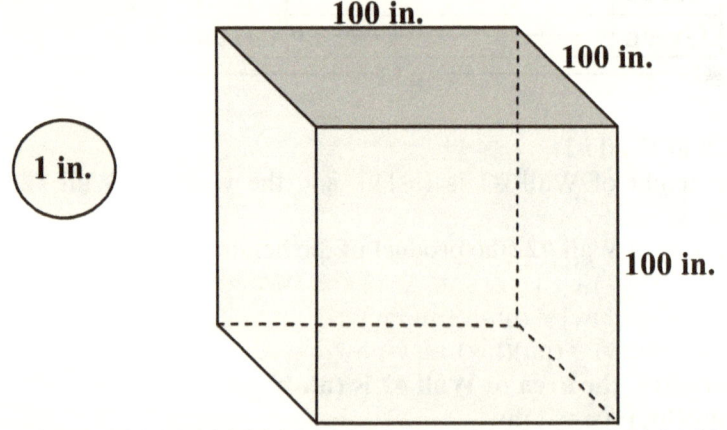

! !

Solution (1):

A trivial solution (<u>not</u> the correct solution, because it is not the maximum number of balls) is as follows:

The bottom layer of balls will consist of 100 x 100 = 10^4 = 10,000 balls. Another 99 layers can be added for a total of :

$$100 \times 100 \times 100 = 10^6 = 1,000,000 \text{ balls}$$

(Reference Diagram 1)

Volume of box:

$10^2 \times 10^2 \times 10^2 = 10^6 \text{ in.}^3$

Volume of balls:

$$\left(10^6\left(\frac{4}{3}\pi\left(\frac{1}{2}\right)^3\right)\right) = 10^6\left(\frac{\pi}{6}\right) = 523,598.8 \text{ in.}^3$$

Volume of one ball = $\dfrac{\pi}{6}$ in.^{3} = .523,598.8 in.^{3}

Unoccupied volume in the box occupied by 10^6 balls:

10^6 in.^{3} – 523,598.8 in.^{3} = 476,401.2 in.^{3}

Percentage of occupied volume:
$$\frac{523{,}598.8}{10^6} = 52.4\%$$

Percentage of unoccupied volume:
$$\frac{476{,}401.2}{10^6} = 47.6\%$$

Solution (2):
The CORRECT solution (maximum number of balls) is as follows:
The balls will not be arranged so that each row will have the same number of balls.
A "staggered" arrangement will more efficiently occupy the cubical volume.
(Reference Diagram 2A).

Each ball of the first layer, and each alternate (odd) layer, is positioned so that it is in the "valley" between any pair of adjacent balls, or in the "valley" between a ball and a side of the box. The first row of these layers will contain 100 balls, the second row will contain 99 balls, the third row will contain 100 balls, the fourth row will contain 99 balls, etc.. The "valley" positioning saves .13397 of an inch for each row (except the first) of each layer.
(Reference Diagram 3).

The number of rows that will fit completely inside the box is determined in the following manner:
The "valley" positioning saves .13397 in. for each row. In other words, each row now requires only .86603 in. instead of 1 in. per row.
Let N = the number of rows attained by the "valley" process.
$$.86603[N]^* = 100 \text{ in.}$$
$$[N] = \frac{100}{.86603}$$
$$[N] = 115 \text{ rows}$$

($*[\ ]$ denotes the "greatest integer" function)

The result is that 115 rows per layer will fit inside the box. The increased number of balls because of the additional rows is greater than the loss of one ball in each alternate row.

The balls of the second layer, and each alternate (even) layer, are positioned in the "valley" configuration. (Reference Diagram 2B). The first row of these layers will contain 99 balls, the second row will contain 100 balls, the third row will contain 99 balls, the fourth row will contain 100 balls, etc.. Again, the "valley" positioning saves .13397 in. for each layer. The number of layers will be 115 by the same calculation that applied to the number of rows in a layer. The result is that 115 layers will fit inside the box so that the top will be able to close completely. The increased number of balls because of the additional layers is greater than the loss of one ball in each alternate row of each layer.

Each of the 58 alternate layers beginning with the first layer (1, 3, 5, 7, ..., 115) will contain 58 rows of 100 balls and 57 rows of 99 balls.

$$58 \times ((58 \times 100) + (57 \times 99)) = 663,694 \text{ balls}$$

Each of the 57 alternate layers beginning with the second layer (2, 4, 6, 8, ..., 114) will contain 58 rows of 99 balls and 57 rows of 100 balls.

$$57 \times ((58 \times 99) + (57 \times 100)) = 652,194 \text{ balls}$$

The total number of balls is therefore:
58 alternate odd layers = 663,694 balls
57 alternate even layers = 652,194 balls
The maximum number is 663,694 + 652,194 = 1,315,888 balls

Volume of balls:

$$(1,315,888)\left(\frac{\pi}{6}\right) = 688,997.3 \text{ in.}^3$$

Percentage of occupied volume:

$$\frac{688,997}{10^6} = 68.9\%$$

Percentage of unoccupied volume:

$$\frac{311,003}{10^6} = 31.1\%$$

The percentage of occupied volume has increased 31.5% from Solution (1) to Solution (2).

The percentage of unoccupied volume has decreased 34.7% from Solution (1) to Solution (2).

The following framed information is not relevant to the solution, but it interestingly presents how the box can be more efficiently packed with cubes having the same volume as the balls. ☺

$e^3 = \dfrac{\pi}{6}$ = volume of a cube that has the same volume as a ball.

$e = \left(\dfrac{\pi}{6}\right)^{\frac{1}{3}}$ = .805996 in. = cube edge.

$\dfrac{10^6 in.^3}{(.805996\, in.)^3}$ = 1,909,859 cubes − each with the same volume as a ball − this many cubes will fit into the box.

$1,909,859\left(\dfrac{\pi}{6}\right)$ = 999,999.834 $in.^3$ (This is close to $10^6 in.^3$)

Box of Balls

Note: The diagram illustrates only one row of the first (bottom) layer of balls. For simplicity and convenience, only ten balls (instead of 100) are illustrated.

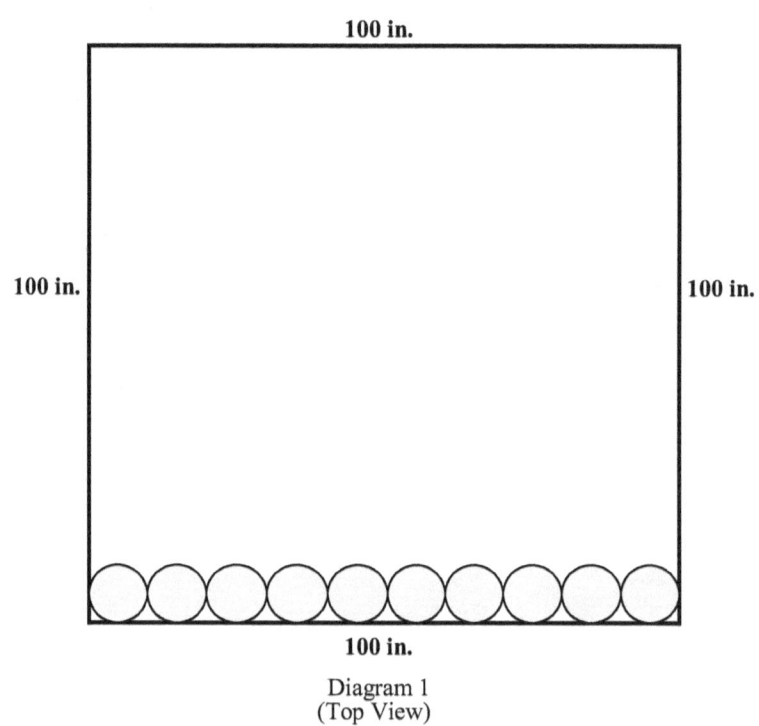

100 in.

100 in.　　**100 in.**

100 in.

Diagram 1
(Top View)

Box of Balls

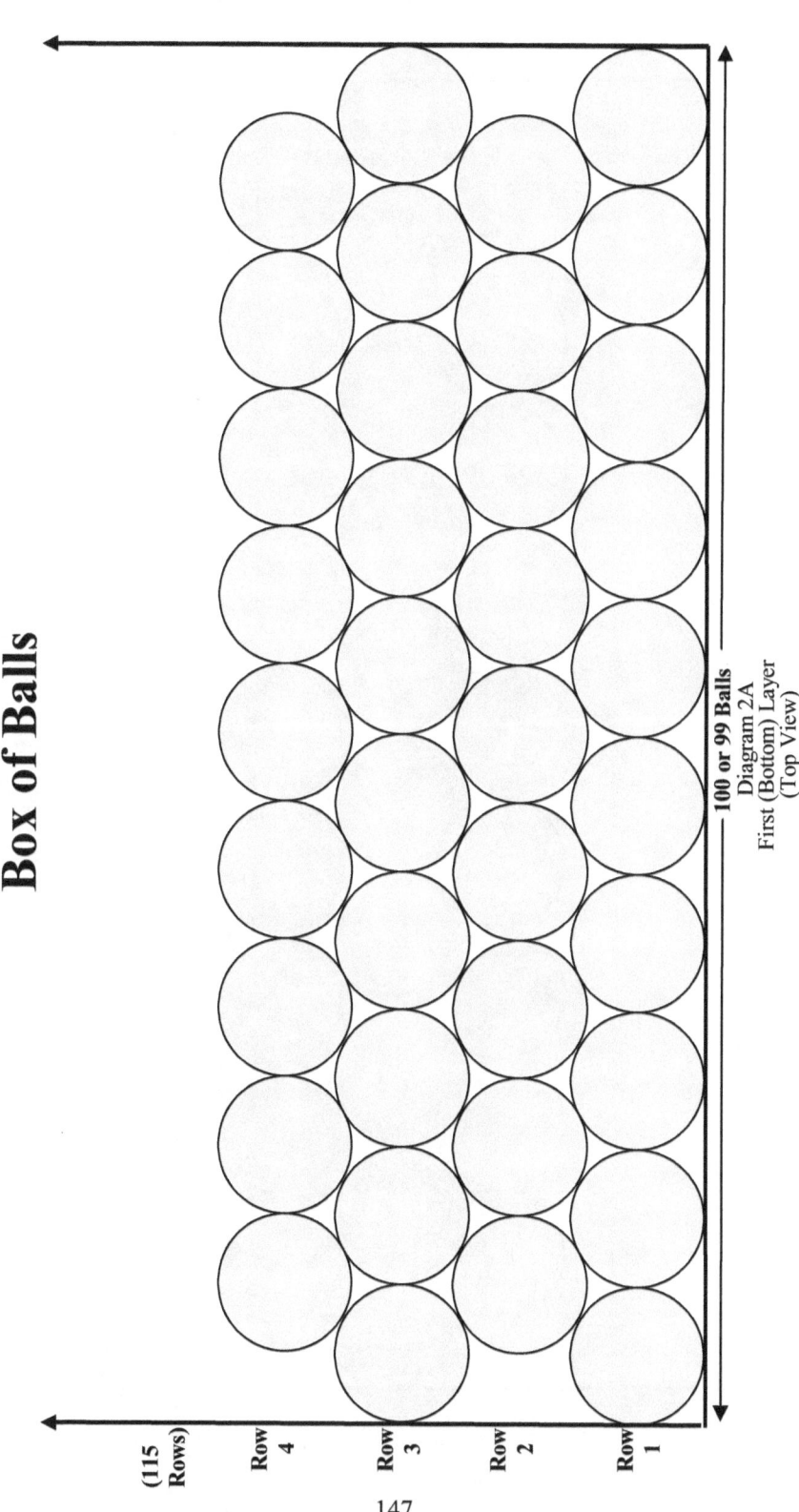

(115 Rows)

Row 4

Row 3

Row 2

Row 1

100 or 99 Balls
Diagram 2A
First (Bottom) Layer
(Top View)

147

Box of Balls

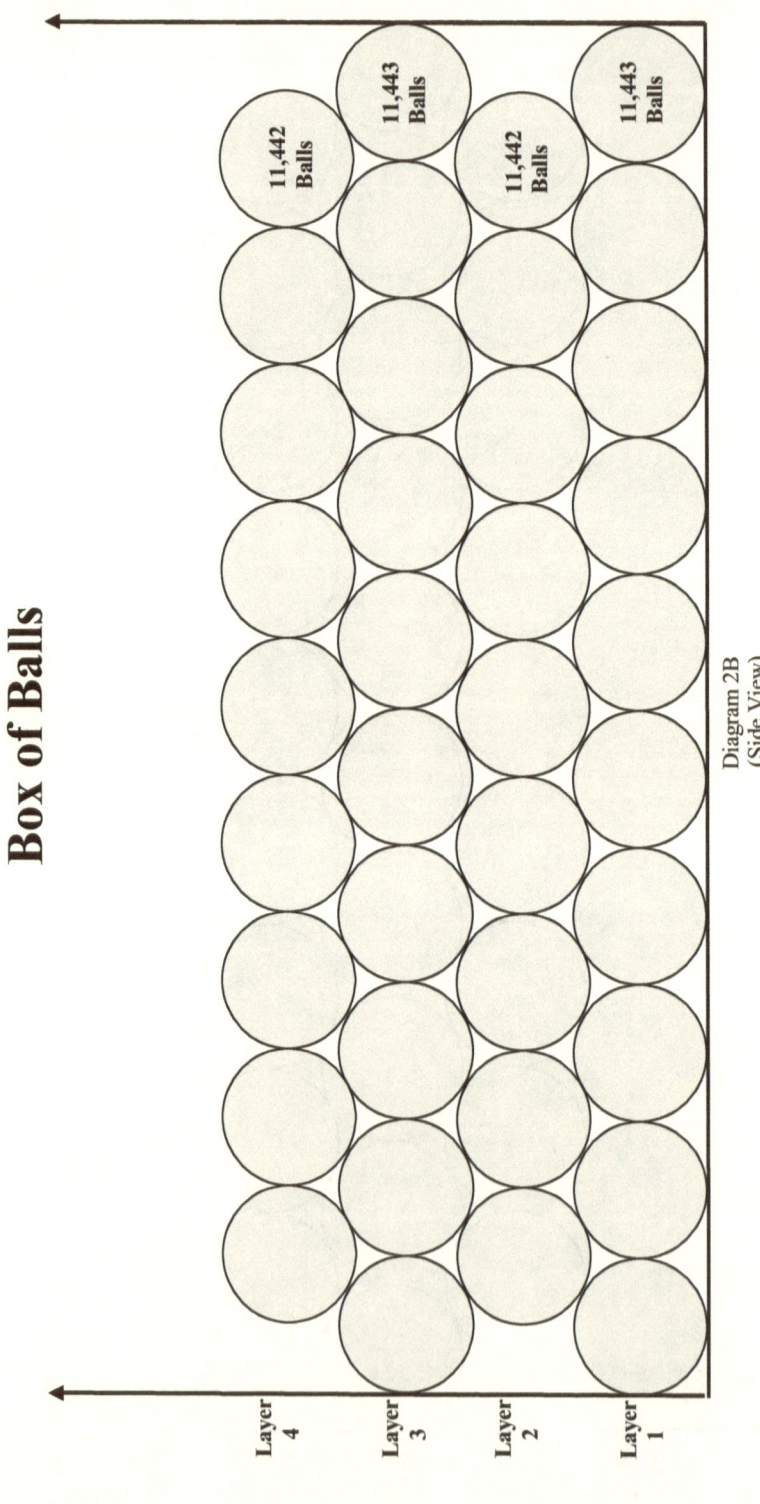

11,442 Balls

11,443 Balls

11,442 Balls

11,443 Balls

Layer 4

Layer 3

Layer 2

Layer 1

Diagram 2B
(Side View)

148

Box of Balls

"Valley Positioning in Rows and Layers"

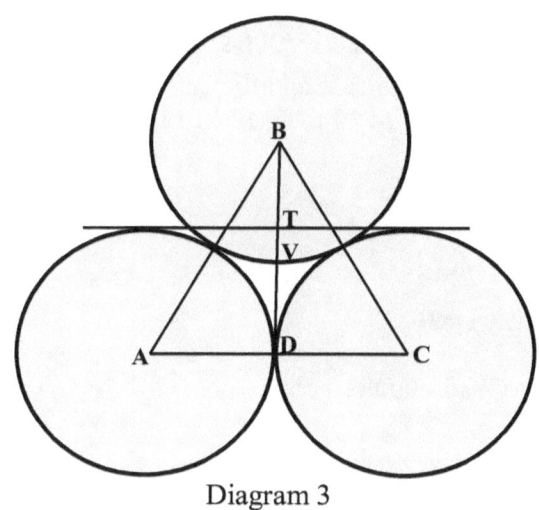

Diagram 3

$$AD = DC = BV = \frac{1}{2} \text{ in.}$$

$$(BD)^2 = (AB)^2 - (AD)^2$$

$$BD = \sqrt{1 - \frac{1}{4}}$$

$$BD = \frac{\sqrt{3}}{2} \text{ in.}$$

$$DV = BD - BV$$

$$DV = \frac{\sqrt{3}}{2} - \frac{1}{2} = .36603 \text{ in.}$$

$$TV = DT - DV$$

$$TV = .5 - .36603$$

$$TV = .13397 \text{ in.}$$

The "valley" positioning saves .13397 in. for each row of each layer, and for each layer.

This indicates that .86603 in. is required for each row and each layer, instead of 1 in.

27
A Number Base Problem

Solve for x:

$324_x = 3064_9$ [324 (base x) = 3064 (base 9)]

(Hint: A quadratic equation is an efficient method of solution.)

!!

Solution:

$3(9^3) + 0(9^2) + 6(9) + 4 = 2245$

$3x^2 + 2x + 4 = 2245$

$3x^2 + 2x - 2241 = 0$

Applying the Quadratic Formula:

$a = 3$

$b = 2$

$c = -2241$

$$x = \frac{-b \pm \sqrt{b^2 - 4ac}}{2a}$$

$$x = \frac{-2 \pm \sqrt{4 - 12(-2241)}}{6}$$

$$x = \frac{-2 \pm \sqrt{26896}}{6}$$

$$x = \frac{-2 \pm 164}{6}$$

$$x = \frac{-2 + 164}{6} \; ; \; x = \frac{-2 - 164}{6}$$

$$x = \frac{162}{6} \; ; \; x = \frac{-166}{6}$$

$x = 27 \; ; \; x = -27\frac{2}{3}$ [Extraneous Root]

$x = 27$

Therefore: $324_{27} = 3064_9$

150

APPENDIX 2
ANSWERS FOR PUZZLES
& RIDDLES – SECTION 2

Puzzles & Riddles

1 Fragility
ANSWER: Silence.
!!

2 Forward & Backward
ANSWER: A Ton.
!!

3 Emotion Malady
ANSWER: The Earth is Bipolar. ☺
!!

4 A Fair-weather Friend
ANSWER: A Shadow.
!!

5 A Convenient Accessory
ANSWER: An Umbrella.
!!

6 A Delicious Word
ANSWER:
By **<u>literally</u>** deleting "SIX LETTERS", the result is
"BANANA".
BsAiNxleAtNteArs
!!

7 A Special Number
ANSWER: The digits zero through nine are in alphabetical
order.
!!

8 Western State Capitals
ANSWER: There are six!
In order from west to east, they are:
Juneau, Alaska
Salem, Oregon
Olympia, Washington
Sacramento, California
Carson City, Nevada
!!

9 Logically a Poor Fit
ANSWER: Shoes are the only "pair" that actually has two
separate pieces.
!!

10 Animal Arithmetic
ANSWER: Chicago!
!!

11 Month After Month
ANSWER: The last three numbers in this series are 3, 3, and 3.
The numbers represent the number of syllables in each of the words for the twelve months.
!!!

12 You Are Looking at the Answer
ANSWER: S: The series consists of the first letter of each word in the question.
!!!
13 This is Difficult and Puzzling
ANSWER: The missing number is 220.
The next number (after 11000) is 11111111111111111111111111

20 is the number 24 written in base 12
22 is the number 24 written in base 11
24 is the number 24 written in base 10
26 is the number 24 written in base 9
30 is the number 24 written in base 8
33 is the number 24 written in base 7
40 is the number 24 written in base 6
44 is the number 24 written in base 5
120 is the number 24 written in base 4
220 is the number 24 written in base 3
11000 is the number 24 written in base 2
111111111111111111111111 is the number 24 in base 1

Our number system is based on 10, but number systems can be based on any other number. The above list defines how to write the number 24 in all the bases from 1 through 12.
!!!
14 A Chemical Compound
ANSWER: H to O, or H_2O.
!!!
15 Irish Eyes are Smiling
ANSWER: Pati O'Furniture
!!!
16 An Alphabetic State of Reference
ANSWER:
There are 26 numbers in this sequence. Each number represents the alphabetic letter in its proper order, and also the number of U.S. States that begin with that letter.
[e.g., there are 4 states that begin with "A".].
!!!

17 A Musical Mixture
ANSWER:
The first two letters of each word (and the pronunciation) are notes of the diatonic scale (music).
!!!
18 A Sneaky Riddle
ANSWER: **Unique** up on him!
!!!
19 Number Fashions
ANSWER:
The 0 said "Nice Belt" to the 8.
!!!
20 Number Fears
ANSWER:
Because 7 8 9 and 10. (7 "ate" 9 and 10)
!!!
21 An Order Perspective
ANSWER:
"The letters of each word in this sentence are in alphabetical order."
!!!
22 The Great Sahara
The Sahara Desert is the world's largest desert (3,500,000 square miles).
ANSWER:
Because of all the sandwiches there!
!!!
23 Homophones
ANSWER:
The listener will assume the "knight" is "night", and obviously will be unable to solve the riddle.
!!!
24 Be Enlightened
ANSWER:
A Candle.
!!!
25 Logical Progression
ANSWER:
The six words are:
Letter, Word, Sentence, Paragraph, Chapter, Book.
!!!

26 A Clever, but Difficult Series
ANSWER:
312211
The line following the answer is:
13112221
Get It?
!!
27 Falling and Breaking
ANSWER:
Daybreak(s) and Nightfall(s).
!!
28 Fore!
ANSWER:
ZERO. If three golfers choose their own balls, the only ball left must belong to the fourth golfer, in which case four (not just three) golfers have their own balls.
!!
29 A Courtly Group
ANSWER:
Tennis Players.
!!
30 An Illustration of Relativity
ANSWER:
Wheeeeeee!
!!
31 A Bucket of Mystery
ANSWER:
A Hole.
!!
32 History/Monroe Doctrine
ANSWER:
"Gentlemen Prefer Blondes".
!!
33 A Corny NFL Riddle
ANSWER:
A Buck An Ear. ☺
!!
34 Sesquipedalianism
ANSWER:
Pneumonoultramicroscopicsilicovolcanoconiosis.
!!

35 Commonality
Kermit and John the Baptist have something in common.
ANSWER:
They both have the same middle name: Kermit the Frog and John the Baptist.
!!!
36 Colors
ANSWER:
Blue Paint.
!!!
37 Buses and Busses
ANSWER:
Buses bring friends closer together; Busses bring close friends even closer together. ☺
!!!
38 Old Testament Facts
ANSWER:
The name of Jeremiah's horse is "Is Me". [Jeremiah said "Woe is me"]. The first of three references is found in Jeremiah 4:31.
!!!
39 A Toothpick Cat
ANSWER:
Solution: Turn the piece of paper upside down so that you now have:

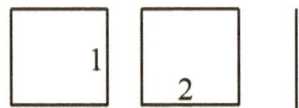

Now move toothpicks 1 and 2 to their new locations to spell CAT.

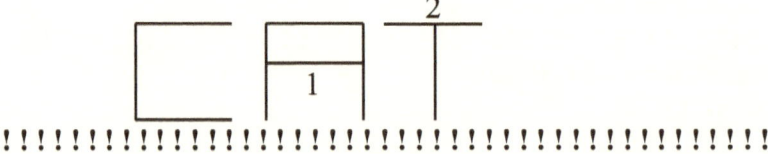

!!!

40 Checkers Puzzle

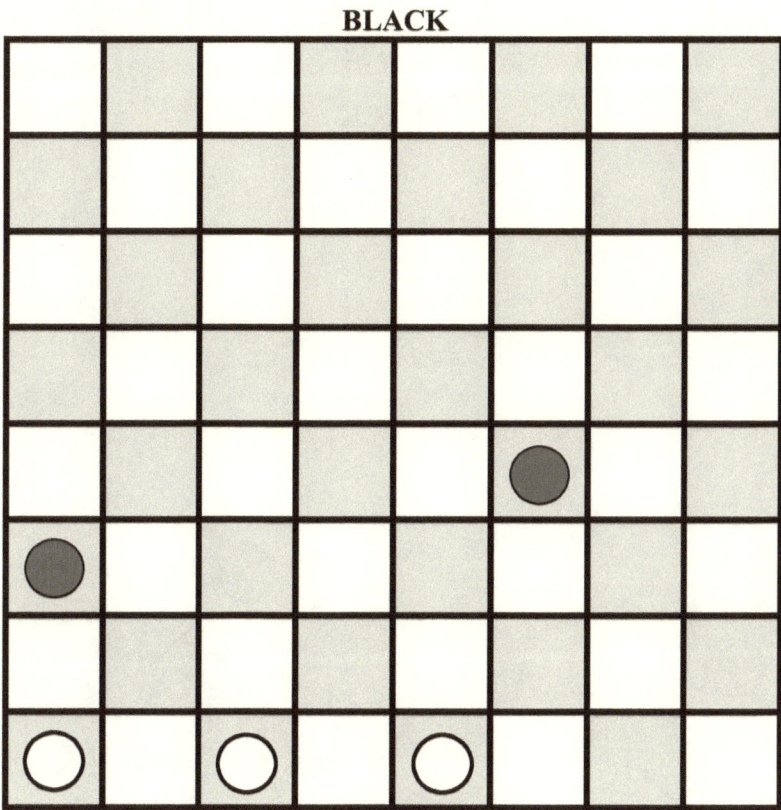

BLACK

WHITE

(Black's Move)

White declares that Black will not be able to crown the piece he moves first.

Is White correct?

Checkers Puzzle

Solution – Move (1)

BLACK

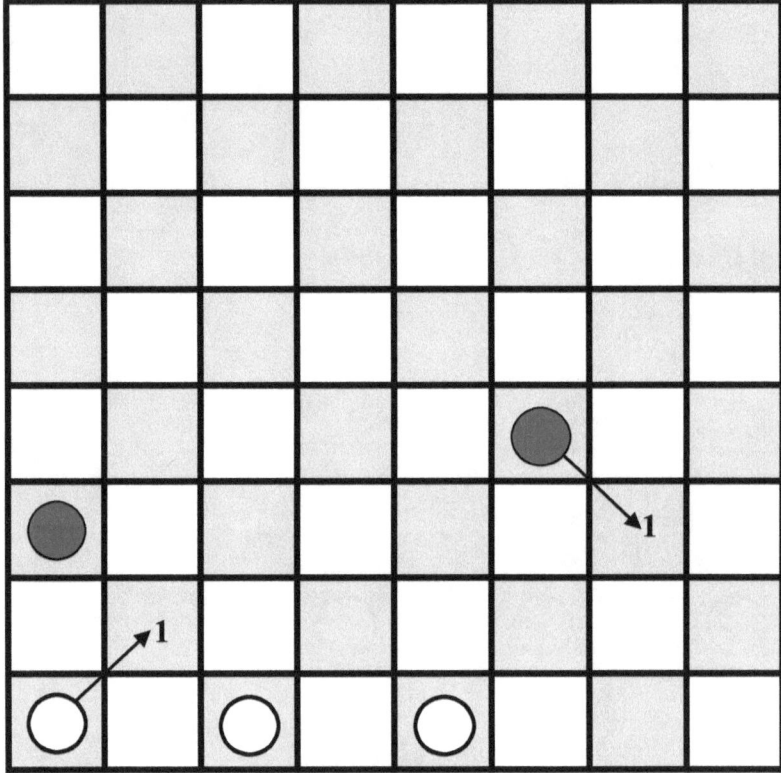

WHITE

Checkers Puzzle

Solution – Move (2)

BLACK

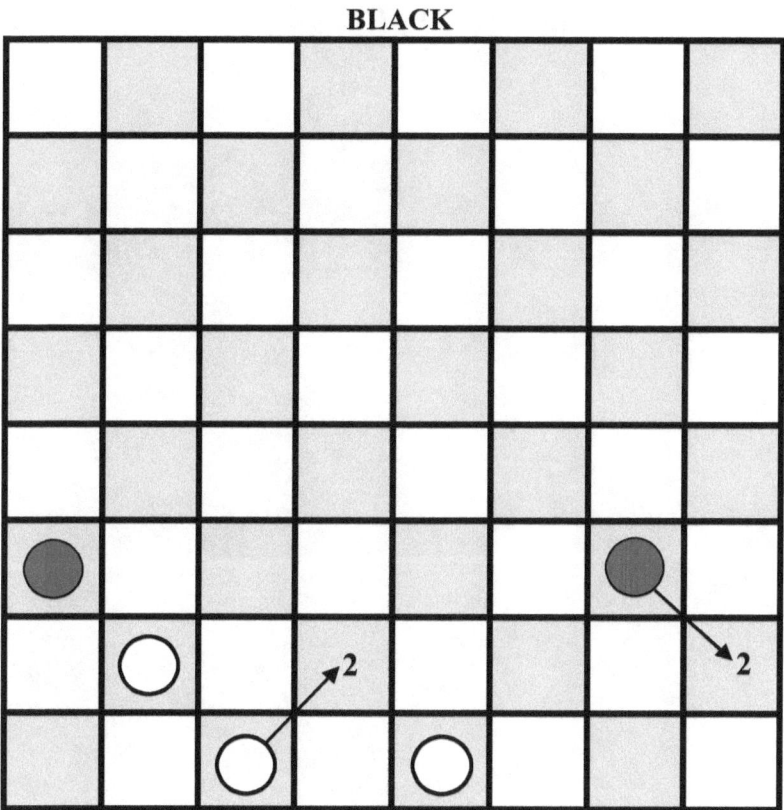

WHITE

Checkers Puzzle

Solution – Move (3)

BLACK

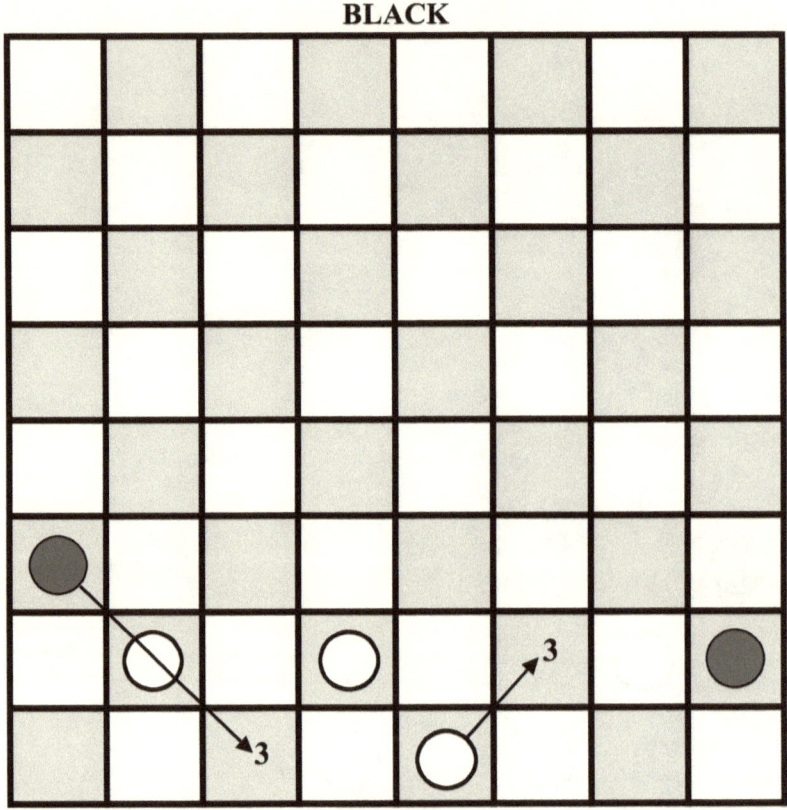

WHITE

Checkers Puzzle

Solution – Move (4)

BLACK

WHITE

Checkers Puzzle

Solution – Concluding Position
White's Declaration is Correct!

BLACK

WHITE
Game Over

This is the first piece moved by Black.
This piece cannot be crowned.

41 A Puzzling Picture
What is it?

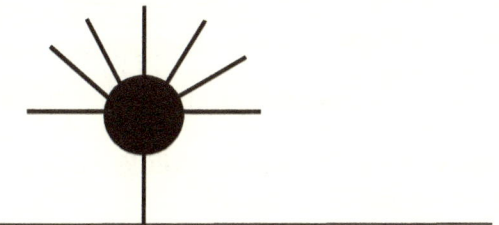

? ?
ANSWER:

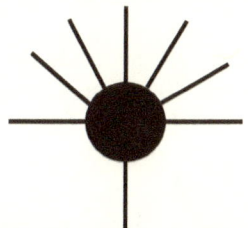

The Spider is doing a Handstand!
! !

42 Twelve to Nine
Answer:
There are twelve toothpicks above the horizontal line of
asterisks.
One is removed.
Now you see only nine.☺

****

****

43 How Many?

Answer:

One solution is to write "one" for each digit. Can you determine a solution using numbers?

!!

44 Sherlock Holmes and the Crimson Snowball

Answer:

Holmes should question Mark Crimson – (? Crimson is "question mark" Crimson).

!!

45 Microwaving

Answer:

The maximum time that can be entered is in the form of 99 minutes, 99 seconds.

Converting 99 minutes, 99 seconds to hours, minutes, and seconds is equal to 1 hour, 40 minutes, and 39 seconds.

!!

46 Pricing Mystery

Answer:

Vowels cost 3 cents, and consonants cost 5 cents. Therefore, Licorice costs 32 cents.

!!

47 An Old Riddle

Answer:

Nothing.

!!

48 A Number Scheme

Answer:

Choose the top number in any column. The product of the three digits is the next number in that column. The same technique is used for all four columns.

!!

49 Relationships

Answer:

There is only **one** invited guest.

Guests (1) thru (5) are the same person. That person is **C**.

!!

50 Geographic Extremes

(1) [48] **North**: Minnesota; **South**: Florida; **East**: Maine; **West**: Washington

(2) [49] **North**: Alaska; **South**: Florida; **East**: Alaska; **West**: Alaska

(3) [50] **North**: Alaska; **South**: Hawaii; **East**: Alaska; **West**: Alaska

Semisopochnoi Island, Alaska is $179.6008°E$.

Amatignak Island, Alaska is $179.1086°W$.

Alaska is <u>farther</u> East than it is West!

!!

51 Spelling

Answer:

The word that is misspelled is "<u>mispelled</u>".

!!

52 Two Answers

Answer:

2,3,4,5,6,7,8,**11**

Spelling is the key: 2 and 3 begin with "T"; 4 and 5 begin with "F"; 6 and 7 begin with "S";

8 begins with "E", and the next number that begins with "E" is 11.

!!

53 Types of People

Answer:

Binary 10 (Base 2) is equal to 2 (Base 10).

Therefore, there are exactly two types of people.

!!

BLACK

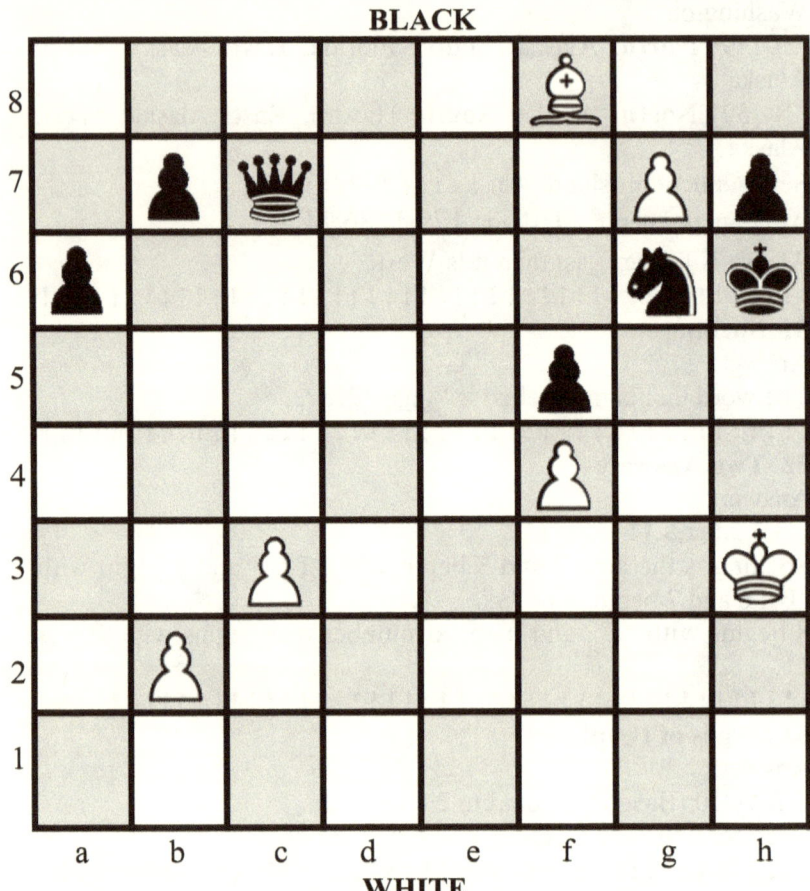

WHITE

(White's Move)
White declares a checkmate in two moves.
Is White correct?

Chess Puzzle

Solution – Move (1)

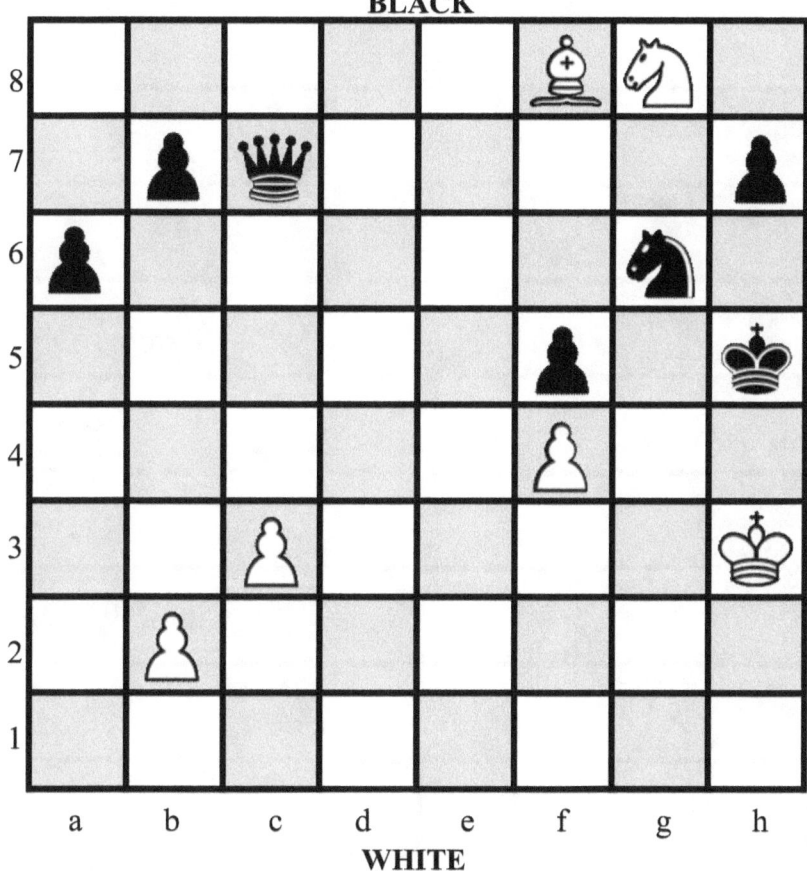

1. g8 = Nch (dbl ch); Kh5

Chess Puzzle

Solution – Move (2)
White's Declaration is Correct!

BLACK

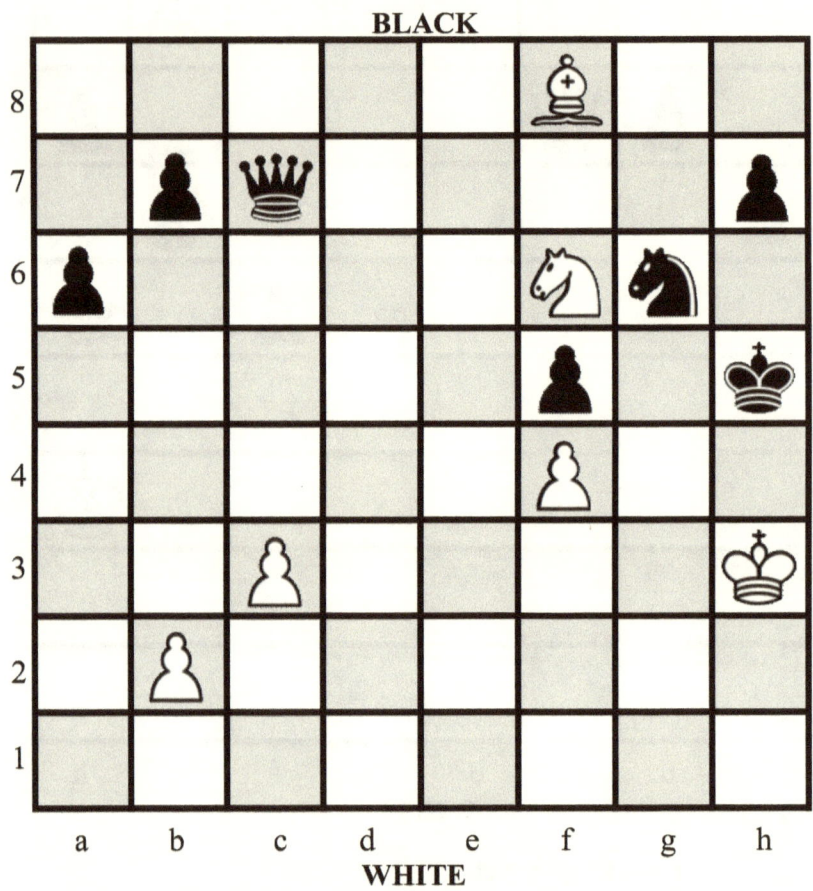

WHITE

1. g8 = Nch (dbl ch); Kh5
2. Nf6 checkmate

!!

55 The Greatest Riddle Of Them All
Answer:
Nothing! Nothing is greater than Jesus!

!!